The Prosperity Revolution

The Non-Technical Guide to the Blockchain

David Conger

Legal Disclaimer
The information in this book is provided for informational purposes only. You use this information completely at your own risk. No part of this book should be taken as legal advice. The author and the publisher are in no way responsible for the consequences you incur when you use this information.

Disclaimer
Many of the sources cited herein are URLs on the Internet. All URLs in this book were checked before publication and found to be valid. However, due to the nature of the Internet, neither the author nor the publisher guarantees that the URLs will remain valid in the future. Both the author and the publisher do hereby certify that all of the sources cited in this book were as accurate and correct as was humanly possible to make them at the time of publication.

Illustrator: David Conger

CONTENTS

Introduction

Imagine you had a special telephone that enabled you to call yourself in 1987, back when the Internet was the quiet purview of a few academics, defense contractors, and government agencies. What would you tell yourself about this new, up-and-coming technology? Maybe you would say, "Invest in Google!" Or perhaps, if your name was Bob, you would tell yourself, "Start an online classified ad site called BobsList!" Perhaps you could direct your younger self, "Start an auction site and call it eBay!"

Whatever instructions you would give to your younger self, it would change your life forever. What if you could **really** be in a situation like that? What if you could find out about a technology that would transform the world **before** it was a big deal? In other words, what if you could find out about it when it was still as unknown as the Internet was in 1987?

Here, I'll make that dream come true for you.

A new technology has been invented that will be as important and impactful as the Internet has been. It's called the blockchain. The blockchain will transform virtually every aspect of our civilization.

You're welcome.

What is this Book About?

This book is about the blockchain. But it is not about the technical aspects of how the blockchain works. In this book, you'll learn what the blockchain is, the basics of how it works, and how it will impact our world.

Chapters 1 and 2 examine the problems we're dealing with in our world. In Chapter 3, we look at why the government is largely useless for solving our problems and how the government should function.

But you may sensibly ask, "What does that have to do with that blockchain-thingy?"

Chapters 1 through 3 help provide the context of how the blockchain can and will be used in the future. Therefore, this book begins by examining the current state of our world and the seemingly insurmountable problems we're facing. It's not an exaggeration to say that we're at a pivotal point in history where we could move into either very dark and turbulent times, or we could create a civilization better than anything that ever came before. The choice is ours.

In Chapter 4, this book provides an overview of the free market and shows how it solves many of the problems that government and big corporations can't. For comparison, this book also takes a look at Marxism (socialism and communism) in Chapter 5 and explains why Marxism is the one thing that will ensure the collapse of our modern world.

With that information, we have enough background to understand liberty and economics in the Age of the Blockchain.

Chapter 6 introduces the idea of a free market of competing currencies where only the most valuable, useful, and stable forms of money survive. You read that right; a free market of competing currencies. As radical as it sounds, you and I *can* create our own forms of money. And we can do a better job of it than the government.

Chapter 7 provides an overview of digital currencies. It talks about the most substantial experiment in digital currencies so far; bitcoin. Chapter 8 then dives into the blockchain and discusses how it can be used to create a free market of digital currencies that are as easy to use and a single, government-issued currency.

In Chapter 9, this book shows how easy and effective it is to use blockchain-based digital currencies. It also provides just a few ideas about how such currencies can be used.

Chapter 10 goes beyond the basics of the blockchain and explains the concept of smart contracts. You'll really like this chapter if you're the type of person that can come up with good ideas for new businesses. Smart contracts are a blockchain-derived technology with a million killer apps. They will start a gold rush like nothing we've seen before.

In Chapter 11, this book introduces the concept of side chains, which are blockchains that work along side of a blockchain-based currency to extend the power of a blockchain even further.

Next, Chapter 12 provides solid examples–some of them based on actual, historical experience–that demonstrate how the blockchain can be applied. It looks at how we can solve massive, seemingly insurmountable social problems without government interference, with little or no new regulations, without taxation, without curtailing our freedoms, and without massive transfers of wealth.

Finally, Chapter 13 examines a few of the many transformative ways that the blockchain can impact business and fulfill the title of this book. It really emphasizes that, with the blockchain, we are only limited by our imaginations in terms of how we structure our economy to make it run better and create new opportunities.

WHO SHOULD READ THIS BOOK?

Here are a few questions that clarify who this book is for.

- Have you heard of the blockchain and want to know what it's all about but you aren't a super-technical programmer-type person?
- Do you want to be in on the Next Big Thing?
- Do you want the chance to invest in hot new companies?
- Do you have good ideas for businesses but don't have the resources to make them a reality?
- Are you interested in the chance to innovate and create something that can be as big as Google, eBay, or Amazon?
- Are you a decision maker in a business that is considering an investment in blockchain technology?
- If you had the opportunity to solve massive problems like income inequality, overconsumption, ecological devastation, poverty, illiteracy, and so forth, would you like the tools to do that?
- If you had the tools to bring the so-called "unbanked" populations of the world into the global economy so that they could lift themselves out of poverty, would you use them?
- If you could restructure our entire power grid in a way that would make a green power revolution profitable, would you want to do that?
- Are you concerned about the concentration of power and wealth from the many to Big Business and Big Government?
- Does it scare you that the so-called 1% have so much control over our lives?

- Do you care about the erosion of your civil rights?
- Do you want the government to just leave you alone and not interfere in your life so much?
- Does liberty matter to you?
- Do you believe in the free market?
- What if you could transcend money entirely and create something better?
- Are you interested in economics and does the idea of custom-designed microeconomies intrigue you?

If you answered yes to any one of these questions, then you should read this book. You don't have to be interested about all of these things. If you care about any one of them, the blockchain provides you with a level of opportunity that you have never had before. It will give you the chance to realized your goals and have a positive impact on the world–or possibly have a positive impact on your own bottom line (which is always nice).

The Prosperity Revolution

The Non-Technical Guide
to the
Blockchain

PART 1

MONEY IN THE REAL WORLD

1 What Is and What Should Be

The world is exactly the way we want it to be. If it weren't, we would make it different.

Given the sorry state of this world, that seems like an astounding statement. But if we look at history, we see that it's true.

For example, there once was a time in the US when it was legal to have separate drinking fountains for blacks and whites. It was legal to force black people to move out of the better seats on city buses.

Then one day, a woman named Rosa Parks just said, "No." She refused to move from her seat on the bus. A young black preacher named Martin Luther King decided that Rosa Parks was right to do what she did. Within a short time, black Americans had organized a citywide boycott of the bus system. And that was a beginning of a mostly peaceful revolution that snowballed until it turned the nation upside down[1].

The US is a very different place now. And it's different simply because Rosa Parks and Martin Luther King decided it should be. They saw the nation as it was. The stood their ground and decided to not give in to what was. Instead, they demanded what should be.

The lesson we can take away from the civil rights movement is that we don't have to settle for what is. We should be courageous enough to look at what is and what should be, and to stand our ground and demand better. We need to have the fortitude to look at massive, institutionalized injustice and say, "No. We will not tolerate this any more."

We live in an age when injustice has been institutionalized on a scale never before seen in human history. To fight against it will take every innovation we

---

[1] Saying that this was a peaceful revolution is not an attempt to diminish the sufferings and deaths of those who were carrying on this fight. Because there *was* violence against them and some *were* murdered. Rather, it is intended to emphasize their forbearance and dedication to not responding with similar violence.

[2] For more information, see Chad Stone, Daniel Trisi, Arloc Sherman, and Brandebot, "A Guide to Statistics on Historical Trends in Income Inequality", Center on Budget and Policy Priorities, February 20, 2015, http://www.cbpp.org/research/poverty-and-inequality/a-guide-to-statistics-on-historical-trends-in-income-

can manage, all of the "out of the box" thinking we are capable of, and all of the courage we can muster.

WHAT IS: POVERTY IS GROWING

Back when I was in Jr. High school, there was a prosperous middle class in the USA. To my young mind, it seemed *everyone* was on the way up. And by "on the way up" I mean that people had the ability to move into higher income brackets. In other words, they had upward mobility in terms of their income and lifestyle.

While this was a very subjective observation at the time, it has become blatantly obvious to everyone that the gap between the rich and the poor is widening[2]. The middle class is struggling and shrinking. I watched this happen over my lifetime and wondered what was causing it.

> "The fact is that this generation -- yours, my generation ... we're the first generation that can look at poverty and disease, look across the ocean to Africa and say with a straight face, we can be the first to end this sort of stupid extreme poverty, where in the world of plenty, a child can die for lack of food in it's belly." - Bono

At the beginning of the 1970s, America had achieved a level of upward mobility that was unrivalled in the history of the world. At that time, the bottom 90% of families received about 67%[3] of the nation's income[4]. The top 1% got about 11%.

But in the mid-1970s all that started to change. By the year 2000, to top 1% of families received 22.5% and the bottom 90% was getting around 50% of all the nation's income[5].

During the period of time from the 1970s to today, the growth of the economy was generally good[6] until the meltdown of 2008-2009. But the expanding economy was leaving an increasing number of people behind. There have been numerous reports and articles attesting to that fact. They

---

[2] For more information, see Chad Stone, Daniel Trisi, Arloc Sherman, and Brandebot, "A Guide to Statistics on Historical Trends in Income Inequality", Center on Budget and Policy Priorities, February 20, 2015, http://www.cbpp.org/research/poverty-and-inequality/a-guide-to-statistics-on-historical-trends-in-income-inequality.

[3] The exact figure, beginning in 1944 is 67.5%, but that figure held roughly steady until the 1970s.

[4] Drew Desilver, "US income inequality, on rise for decades, is now highest since 1928", Pew Research Center, December 5, 2013, http://www.pewresearch.org/fact-tank/2013/12/05/u-s-income-inequality-on-rise-for-decades-is-now-highest-since-1928/.

[5] Ibid.

[6] Yes, there were definite ups and downs during that time period. But the 2008-2009 meltdown was something unlike the downturns that had occurred before.

blaze such titles as "Growth Has Been Good for Decades. So Why Hasn't Poverty Declined?"[7] and "Men's Earnings Haven't Grown Since the 1970s— Why?"[8] but they provide no real answers.

In neighborhoods that have always been poor, things are getting very, very bad. The quality of life, which was already not good, is declining rapidly. By any objective measure, these neighborhoods are experiencing higher crime rates, failing schools, and far fewer job opportunities[9].

> "The life of the community, both domestically and internationally, clearly demonstrates that respect for rights, and the guarantees that follow from them, are measures of the common good that serve to evaluate the relationship between justice and injustice, development and poverty, security and conflict." - Pope Benedict XVI

But that's not the whole story. Poverty is spreading to suburban areas of American that have long been otherwise prosperous. In just the 11 years between 2000 and 2011, poverty in the suburbs grew by an astounding 64%![10]

With the growth of poverty, there have been increasing calls for more government programs. And the government has responded willingly. We now live in a day when 49.2% of all American households[11] are receiving some sort of financial assistance from the federal government. That's just not sustainable. Something has to give.

> "In the 21st century, I think the heroes will be the people who will improve the quality of life, fight poverty and introduce more sustainability." - Bertrand Piccard

We see the results of the growth of poverty in our news broadcasts and headlines. The Occupy Wall Street protests, as ill-conceived as they were, are a perfect example. It mirrors popular uprisings of many previous centuries. One only has to read the histories of the French Revolution to see where this is going (hint: guillotine).

---

[7] "Growth Has Been Good for Decades. So Why Hasn't Poverty Declined?", New York Times, June 6, 2014, www.nytimes.com/ 2014/ 06/ 05/ upshot/ growth-has-been-good-for-decades-so-why-hasnt-**poverty**-declined.html.

[8] Derek Thompson, "Men's Earnings Haven't Grown Since the 1970s—Why?", The Atlantic, September 12, 2012, http://www.theatlantic.com/business/archive/2012/09/mens-earnings-havent-grown-since-the-1970s-why/262296/.

[9] The Federal Reserve System and the Brookings Institution, "The Enduring Challenge of Concentrated Poverty in America: Case Studies from Communities Across the US" (Richmond, VA: 2008); and Patrick Starkey, Stuck in Place: Urban Neighborhoods and the End of Progress Toward Racial Equality (Chicago: University of Chicago Press, 2013)

[10] Elizabeth Kneebone and Alan Berube, Confronting Suburban Poverty in America, The Brookings Institution Press, January 13, 2014, see Chapter 2 in its entirety.

[11] "We've Crossed The Tipping Point; Most Americans Now Receive Government Benefits" Forbes, July 2, 2014, http://www.forbes.com/sites/merrillmatthews/2014/07/02/weve-crossed-the-tipping-point-most-americans-now-receive-government-benefits/.

WHAT SHOULD BE: A PATH TO PROSPERITY

In a real, functional economy in which everyone had a real chance at success, there would be a path out of poverty for virtually everyone who was willing to work and educate themselves and their children.

> "The first panacea for a mismanaged nation is inflation of the currency; the second is war. Both bring a temporary prosperity; both bring a permanent ruin. But both are the refuge of political and economic opportunists." - Ernest Hemingway

Impossible you say? But wait, *that's the American Dream!* This is what we *used* to believe about America. There was a time when most people believed that if you just work hard and get a good education, or learn a solid trade, you could move up in the world and your kids could reach a better standard of living. Have we become so cynical about the American Dream that we no longer believe it's possible to attain?

We don't have to settle for what is. We really can achieve what should be. There should be and can be a way to rise from poverty for those who will make the effort. It's as simple as that.

Laws in this country should not favor the rich and Big Business. For instance, there should be no such thing as the Monsanto Protection Act, a bill that provides special benefits to the gigantic Monsanto Corporation and was embedded inside an even larger bill so it wouldn't really be noticed. If Monsanto gets protection from Congress, why shouldn't I? Shouldn't we all be equal before the law? Isn't that the basis of democracy?

When the government picks winners, everyone loses. The government should not be allowed to play favorites when it comes to the economy (or anything else for that matter). When Congress passes the Monsanto Protection Act, or any other bill that favors a particular business or group of businesspeople, it is simply corruption. The congressmen and congresswomen who introduced and voted for the bill should be thrown out of office and should go to jail.

Likewise, cities and states that offer tax advantages or other incentives to big companies to move into their areas are suborning democracy. I don't get those advantages and incentives. It doesn't matter that they're big and I'm small. I have the same rights before the law as anyone in large corporations like Microsoft, Intel, Apple, and Google, all of which have received special favors from cities and states that I haven't.

When a businessman gives a politician money or special favors in exchange for extra consideration or special advantages, we call that corruption and bribery, and we put both the businessman and the politician in jail. But when it operates in reverse, and the politician gives the businessman money and special favors we say that's just good politics and we reelect the politician.

> "The point in history at which we stand is full of promise and danger. The world will either move forward toward unity and widely shared prosperity - or it will move apart." - Franklin D. Roosevelt

No. It's just as corrupt. It's bribery. Letting the politician bribe the businessman is just as bad and just as illegal as letting the businessman bribe the politician. But we not only ignore it, we cheer it on. There is no sense in this at all. We are destroying our democracy.

If we are to have a system in which all are equal before the law, the government can't grant any favors to particular businesses and not grant those favors to another. My business is just as important to me as Monsanto's is to it. My business deserves the same consideration as Monsanto's even if I only employ myself or just a few people. That's what it means to be equal before the law. Under the rule of law, all laws apply equally to everyone.

Giving some business, professions, or groups special economic advantages kills the chances of others to rise from poverty. In a world where only the well connected can get ahead, there is no hope for the

> "We begin to change the world when we stimulate long-term prosperity using technology. There is not a problem that's large enough that innovation and entrepreneurship can't solve." - Naveen Jain

growing number of people who are sinking below the poverty level. In a world where we are all equal before the law and have an equal footing in the economy, everyone has a viable path out of poverty.

We stand at a junction in history. A new technology, called *the blockchain*, has been invented that will impact the world at least as much as the Internet did. It's true. The blockchain can enable us to completely restructure our government and our economy in ways that are more free, more fair, and that offer real paths out of poverty.

WHAT IS: THE WORLD IS RUNNING OUT OF MONEY

As nutty as it seems, the world is running out of money. Everywhere, nations are falling into one economic crisis after another. Here are some examples.

THE UNITED STATED OF AMERICA

In 2008, a massive bubble in the US housing market burst. Banks had lent out enormous amounts of money for mortgages that people could not pay back. When it became clear that this was the case, banks large and small across America began to fail[12].

> "Bubbles arise if the price far exceeds the asset's fundamental value, to the point that no plausible future income scenario can justify the price." — Justin Fox

The remaining banks tightened their lending policies. Not only home loans, but consumer loans, business loans, and loans for short term liquidity became difficult or impossible to get.

Unemployment was the result–along with economic stagnation, losses in household income, and sharp price jumps all through the economy. This quickly spread across the globe and turned into the worst financial crisis since the Great Depression.

The hit to financial markets has resulted in a decline of the real economy throughout the whole world[13]. The US economy in particular and the global economy in general have not yet fully recovered from the meltdown. Recession is increasingly seen as the "new normal" for our economy. The money just seems to be gone.

GREECE

In October of 2009, the newly-elected government in Greece revealed what the previous government had diligently kept secret. That is, the country's debt was actually twice what had been reported[14]. People were astounded to learn that the previous deficit figures of 6% of GDP was an outright lie. It was actually 12.7%.

Markets around the world reacted with an immediate and dramatic loss of confidence in Greek's ability to pay its debts. That sparked a massive downturn in the Greek economy, unemployment, power outages, riots, and harsh austerity measures.

---

[12] Bulent Gokay, "The 2008 World Economic Crisis: Global Shifts and Faultlines", The Centre for Research on Globalization, February 15, 2009, http://www.globalresearch.ca/the-2008-world-economic-crisis-global-shifts-and-faultlines/12283

[13] Ibid.

[14] "The Greek Crisis Explained", France24, March 10, 2010, http://www.france24.com/en/20100504-greek-crisis-eu-imf-bailout-debt-economy-germany-france-unions-jobs-euro-markets.

Since then, the other nations of the Eurozone have had to give multiple, very large bailouts to Greece to keep it from running out of money. They fear that it will collapse completely and drag down the rest of the Eurozone with it.

CYPRUS

In 2013, the government of Cyprus limited the amount of cash people could withdraw from their bank accounts. They then stated that if you had more than 100,000 euros in the bank, they were taking half[15].

Yes, that's right. The government just seized everyone's money.

The Crypriot bank, Popular Bank, turned over the money as directed by government officials. Much of that money had been deposited by Russian mobsters. So people didn't cry too much. But of course, Russian mobsters don't tend to be too nice about that sort of thing.

Anyway, the rest of the money was held by the common citizens of Cyprus. The government just impoverished a good portion of its population in one fell swoop.

And of course, Popular Bank couldn't survive such a devastating blow to its assets. When it looked like it would fail, the government nationalized it.

Why did all of this happen?

Greece happened. Yes, that whole Greek economic collapse thing.

Cyprus, which is closely affiliated with Greece, was hit heavily by the failure of the Greek economy. Its government and its banks were very exposed to the massive drop in the value of the assets of Greek banks. When Greece ran out of money, so did Cyprus.

SPAIN

Like Greece, Spain has run out of money. Unemployment has risen as high 25%[16] and currently sits at around 23%[17] as of this writing. There have been a

---

[15] Dylan Matthews, "Everything you need to know about the Cyprus bailout, in one FAQ", The Washington Post, March 18, 2013, http://www.washingtonpost.com/blogs/wonkblog/wp/2013/03/18/everything-you-need-to-know-about-the-cyprus-bailout-in-one-faq-2/

wave of mortgage foreclosures, some of which have resulted in suicides. For example, one woman felt so heavily hit by the foreclosure on her longtime family home that she actually set herself on fire in a bank[18].

Spain has since made some good progress on a recovery, but it's still questionable whether or not the crisis in Spain will stretch the finances of the Eurozone so thin that it could cause the 17-nation coalition to crumble.

And More

I could continue to regale you with tales of the many recent economic collapses that have occurred around the world. In recent years, country after country has moved to the edge of an economic abyss. And experts seem to be at a loss to explain it or to remedy it.

WHAT SHOULD BE: A WORLD OF FINANCIAL STABILITY

Most economists would agree that there is a level of economic activity at which the economy could potentially stay forever. They call this state *full employment* and it occurs when there is a level of production in which all of the inputs to production, such as the machines used to produce things, are being used. But they are not being used so extensively that they wear out, break down, and so forth. If nothing disturbs the economy, such as a war or natural disaster, the state of full employment can be maintained forever.

But we don't have that now. Instead, we have what is euphemistically to as "the business cycle." Unfortunately it is not a cycle and it doesn't really have anything to do with business.

The business cycle is a series of market bubbles that grow up, causing certain parts of the market to be overinflated, and then burst, causing the *entire* market to shrink. These unending market bubbles result in booms and busts that get bigger each time we go through this nightmare of a merry-go-round.

[16] Ciaran Giles, "Spain Unemployment: 1 In 4 Out Of Work", Huffington Post, October 26, 2012, http://www.huffingtonpost.com/2012/10/26/spain-unemployment_n_2025328.html
[17] "Spain Unemployment Rate 1976-2015", Trading Economics, http://www.tradingeconomics.com/spain/unemployment-rate
[18] "Spanish Woman Self-Immolates In Bank", The Huffington Post, February 18, 2013, http://www.huffingtonpost.com/2013/02/18/spanish-woman-self-immolates0bank_n_2713131.html.

It eats away at our savings, diminishes our earning power, and threatens the global economy itself.

However, we don't have to put up with this at all. Under a truly free market, business cycles are completely unnecessary. While it's true that there will always be some ups and downs in the market, the business cycle is something that we don't have to have[19]. Barring severe disasters, the market should proceed in a much more steady state. And in fact it can. We can build economic structures that end the business cycle and give us a stable, sustainable, and viable world economy. Again, the blockchain enables this and democratizes the economy like never before. We'll discuss this in detail in later chapters.

WHAT IS: GOVERNMENT IS GROWING LIKE A CANCER

When I was in Jr. High school, my US history teacher explained the checks and balances on government in the US Constitution. He talked about how the balance of power in the government can sometimes shift over time. Then he mentioned that the balance of power had bee steadily shifting toward the executive branch since the early part of the 20th Century[20].

That statement caught my interest, so I raised my hand and asked, "Why?"

His response was surprising to me. "I don't know," he said. "But it's been a steady process for the last 60 or 70 years."

> "The makers of the Constitution conferred, as against the government, the Right to be let alone; the most comprehensive of rights, and the right most valued by civilized men." - Louis D. Brandeis

I have remembered this incident all of my life. And for years I continued to wonder what was causing this power shift as I watched it play out over the remainder of the 20th Century and into the new millennium. Here are just a few of innumerable examples that I could provide. But they definitely serve to illustrate the current march toward folly that the government is trudging forward with.

---

[19] Christina D. Romer, "Business Cycles" The Library of Economics and Liberty, http://www.econlib.org/library/Enc/BusinessCycles.html.
[20] This conversation occurred in about 1972.

BUREAUCRATIC BRAIN DEATH

Government control over all aspects of human life is growing on an unprecedented scale. Lately, it seems that the government has gone crazy. They're doing things that make no sense to anyone with a brain. For example, state run elementary schools suspend their very young students for eating a Pop-Tart into a shape that vaguely resembles a pistol. This actually happened. And more ridiculously, the school district actually needed a hearing to determine whether this was a valid action or not. And astoundingly, they upheld the suspension[21].

Likewise, an Oregon man was sentenced to 30 days in jail for collecting rainwater for his garden because the government thinks that it owns all the water in the state[22]. This, of course, flies in the face of thousands of years of both tradition and law. And it means that Oregon bureaucrats have complete control over people's lives. Their decisions can cut people off from one of the most basic substances they need from their survival and there's nothing they can do about it. If they attempt to provide themselves with the water that falls naturally on their own property that they have bought and paid for, the government will lock them up.

What part of this sounds like we're in a free country?

CENTRAL CONTROL OVER ENERGY

For generations now, both the states and the federal government have done their best to control the production and sale of energy in this country. It is exactly this fact that keeps green energy from flourishing.

> "Manufacturing and commercial monopolies owe their origin not to a tendency imminent in a capitalist economy but to governmental interventionist policy directed against free trade." - Ludwig von Mises

Although some states give subsidies, usually in the form of tax credits, for homeowners who install solar or wind generators, every state vigorously protects the power companies' monopolies on the sale of electricity. There are huge regulatory barriers to starting a power company. So for example, if someone with a

---

[21] "School rules against Md. second-grader suspended for gun-shaped Pop-Tart", The Washington Times, June 13, 2013, http://www.washingtontimes.com/news/2013/jun/13/school-rules-against-second-grader-suspended-gun-s/.

[22] "Man jailed for collecting rainwater in illegal reservoirs on his property", Fox News, August 16, 2012, http://www.foxnews.com/leisure/2012/08/16/man-jailed-for-collecting-rainwater-in-illegal-reservoirs-on-his-property/.

wind generator produces more power than she needs, it's almost impossible for her to sell her extra electrical capacity to her neighbors.

Suppose you live in a sunny place and have enough solar panels to produce more power than you use. If you can't sell your power to your neighbors, the sensible thing would be to sell your excess electricity to the power company, right?

Wrong, they don't want it. And they'll do their best to not pay for it if you do provide power for them. They certainly won't pay anything like a fair market price.

If we had a reasonable system in place, the power companies would have to pay you about 50% of what they sell their electricity for. When Target or Wal-Mart sells stuff, the price they pay their suppliers is about 50% of what they sell it for. Why should electricity be any different?

But because electricity is a government-granted monopoly, the power company has no competition. And this gives them complete control over the market. In a very different sense than is usually intended, they really are the power company (pun intended).

The regulations controlling power companies let them get away with things that almost no other business can get away with. They can pay far less than the market value for electricity. This virtually ensures that small scale electrical producers can't get a foothold in the market. And of course, that's exactly what they want.

And all of this doesn't even begin to touch what's happening with the government's war on coal. Rules for power plants imposed by the EPA make it nearly impossible to build a new coal power plant[23].

Here again, the government is picking winners and losers. While short-sighted bureaucrats may crow about all of the pollution they've kept out of the air by shutting down coal plants, the government is actually cutting us off from a green future.

> "We need an equal opportunity society, one in which government does not see its job as picking winners and losers. Where do you go if you want special favors? Government. Where do you go if you want a tax break? Government. Where do you go if you want a handout? Government. This must stop." - Bobby Jindal

As the coal power plants close down, electrical prices go up. Under normal market conditions,

---

[23] John E. Sununu, "Obama's war on coal", The Pittsburgh Post-Gazette, October 15,2014, http://www.post-gazette.com/opinion/2014/10/15/Obama-s-war-on-coal/stories/201410150036.

a price increase would cause investors to rush to build new power plants. But when the market is as heavily regulated as the energy market is, investors are hesitant to put money into the development of power production. Seeing the government kill off coal has to make them think twice about such investments. Who would want to build a wind farm when there's a strong possibility that the government may decide at some time in the future that wind generators kill too many birds? Why would anyone in their right mind put up expensive wind generators if, in a few years, the government might come along and regulate them out of business?

Green energy will never become commercially viable for as long as there is central control over the sale and production of electrical energy.

CENTRAL CONTROL OVER THE INTERNET

Recently, the FCC has decided that the Internet is a public utility rather than a collection of private networks. It compares the Internet to the power system.

> " [With FCC regulation of the internet] we'll end up with an Internet that is more regulated, more subject to regulatory uncertainty in the near-term, and more like a public utility from another era than an information delivery service for the modern age. It'll be 2015—but for the Internet, it'll be 1934 all over again." – Paul Suderman

Unfortunately for the FCC, they're idiots. If the Internet is comparable to anything, it's the printing press. The Internet is a collection of privately-owned computers that are used to publish information (just like printing presses). These computers are connected over multiple privately-owned data networks. In no way does this resemble a public utility like the power and water grids.

Currently, the FCC says it will not regulate Internet content. But the regulations actually leave that up to the commissioners that control the FCC. Even if the current commissioners never exert their power to control who

> "I am disturbed by how states abuse laws on Internet access. I am concerned that surveillance programmes are becoming too aggressive. I understand that national security and criminal activity may justify some exceptional and narrowly-tailored use of surveillance. But that is all the more reason to safeguard human rights and fundamental freedoms." - Ban Ki-moon

can access the Internet and who can publish information on it, does anyone actually think that no commissioners ever will? If your answer is yes, you are singularly naïve.

If the FCC is allowed to continue this unconstitutional power grab, what will they take control of next? This precedent means they can now regulate any means of communication. Does that sound like free speech? Is there any practical limits to their power? And will they target unpopular speech the way the IRS recently targeted the opponents of the Obama administration?

The FCC was created to auction off access to the limited electromagnetic spectrum, nothing more. Now it controls every form of communication imaginable—except maybe smoke signals (but we have the EPA to regulate that).

CENTRAL CONTROL OVER EDUCATION

In 2014, Barak Obama quietly issued a directive to the Department of Education that was roundly ignored by the mainstream press. This directive mandates that schools, both public and private, conform to federal education standards in order to receive funds.

These standards enable the federal government to bypass the limitations of the already intrusive Common Core. They enable the federal government to take control over all aspects of teacher training and dictate both what is taught in schools and how it is taught.

This backdoor directive gives the federal government a powerful tool over states that are reluctant to adopt Common Core. Although the federal government repeatedly insists that Common Core is a "state-led" effort that is "strictly voluntary," it is using every means possible to force states to adopt it.

And no matter how many times they repeat that the states are in control of Common Core, it is the Department of Education, not the states, that oversees the assessment test design for the Common Core national standards[24]. Because schools

> "When the state or federal government control the education of all of our children, they have the dangerous and illegitimate monopoly to control and influence the thought process of our citizens." - Michael Badnarik

under Common Core are allocated their money based in part on the results of the standardized tests, control over the tests is control over classroom content. The teachers, of course, will teach to the tests in order to ensure their funding. Simply put, the adoption of Common Core enables Washington D.C. to dictate what is taught and how it is taught in every classroom in America.

---

[24] Carherine Gewrtz, "Common-Assessment Groups to Undergo New Federal Review Process", Education Week, April 1, 2013,
http://blogs.edweek.org/edweek/curriculum/2013/04/common_assessment_groups_to_undergo_new_feder al_review_process.html

BIGGER AND BIGGER

The government is the last growth industry left. The percentage of government employees (GE) as a percentage of the population (GE/P) has seen steady growth (about 7%) since the 1980s that is consistently above the growth rate of nearly any other economic sector[25]. The federal government alone now employs over 2 million people[26]. This, of course, doesn't even begin to count all of the state, city, and country employees throughout the nation.

You might object that we need that many employees to provide the services that we all use.

False. And I'll demonstrate how and why throughout the rest of this book. It's precisely because of the blockchain that most government services can be provided without the government being involved at all, as we'll see in later chapters. And I'll show how using government to provide these services actually makes us less prosperous, increases the number of people who fall into poverty, and makes it harder to climb out of poverty.

For now, it's enough to understand that there is only so much money to circulate in the economy. As the government grows in size, it consumes more of that money. Yes, the money it spends on its unheralded number of employees does get back into the economy. But even so, it doesn't stimulate the economy at all.

Say what?

"Every year the Federal Government wastes billions of dollars as a result of overpayments of government agencies, misuse of government credit cards, abuse of the Federal entitlement programs, and the mismanagement of the Federal bureaucracy." - Chris Chocola

When private industry spends money, it spends capital in the most efficient way possible. It tries to get the absolute most for its money. If it doesn't succeed, it can go out of business.

Government, on the other hand, is under no such imperative. It can waste money in any way it wants. I remember as far back as the 1970s reading news articles about the Department of Defense paying

---

[25] Mike Patton, "The Growth Of Government: 1980 To 2012", Forbes, Janurary 24, 2013, http://www.forbes.com/sites/mikepatton/2013/01/24/the-growth-of-the-federal-government-1980-to-2012/.

[26] Stephen Dinan, "Largest-ever federal payroll to hit 2.15 million", The Washington Times, February 2, 2010, http://www.washingtontimes.com/news/2010/feb/02/burgeoning-federal-payroll-signals-return-of-big-g/?page=all.

$1,500 for an ordinary hammer. That was the first time I heard about government waste. Since then, I've seen a veritable onslaught of reports on how the government squanders our money. And do I really need to provide references on this? It's easy to just do an Internet search on "government waste" to find any amount of proof of this that you want.

The point is that because the government doesn't spend money in efficient and effective ways, it causes the economy to produce the wrong things. If government officials don't have to worry about spending $10,000 for their office chair, then companies will gladly produce more high-end office chairs.

While this might sound great for office chair companies, that money *isn't* being spent on other things—essential things that everyone needs and uses. Massive government waste ties up capitol (money that can be invested) that would otherwise go to creating new businesses and jobs. The money that is redirected away from being invested in innovation ends up being a drag on the economy because it's being spent in the wrong places. The economy doesn't produce the right goods and services so it functions less efficiently. And therefore it produces fewer jobs than it otherwise would.

And of course there are only so many workers in our nation. People who work in government *are not* working in industry. That results in fewer goods and services produced in the economy and less innovation.

In other words, the more people work in government—and the more money we spend on government employees—the fewer jobs there are in the overall economy. The economy would actually produce *more* jobs if the government wasn't so large. A growing government means a shrinking private sector. Or at least a private sector that is not as large as it could be.

Efficient investment in the private sector produces far more jobs than inefficient "investment" in the government[27]. As the government grows larger, we limit our own economic future. We limit the number of jobs we produce. We limit the amount of money we can invest in new ideas. We even limit the amount of money we can spend on maintaining our existing infrastructure. In short, we are squandering our future. This is true even if we use deficit spending to flood the economy with money we don't have. All it does is make things harder later. And someday, the piper *must* be paid.

---

[27] Government can't really "invest" in anything. They just use the word invest to make it seem like they're doing the same thing that businesses do when they create new products, services, markets, and jobs. The government can do exactly *none* of those things.

THE ULTIMATE CONTROL

In 2013, an economist named Larry Summers delivered a speech[28] at the International Monetary Fund Research Conference that should scare the bejeebers out of everyone. Our friend Mr. Summers wants to eliminate cash.

Why is that bad, you say? Everyone uses credit and debit cards anyway, you say?

To answer that, we have to take a look at the world's current financial situation.

It stinks.

We Spent It All

To be more specific, governments have spent all the money they could get their hands on like a bunch of drunken sailors on shore leave.

No. I shouldn't say that. It slanders drunken sailors.

In any case, due to astonishingly excessive and stupid government spending, layer after layer of incredibly moronic government regulations, a worshipful devotion to brainless and dysfunctional Keynesian economics, and the advance of mindless dogmas such as socialism and communism, the governments of the world are rapidly bankrupting themselves.

The Summers Solution: Take More Money from You

To keep the gravy train going, and to avoid the massive worldwide chaos that would ensue in the event of a large-scale economic collapse, governments are looking for more ways to take your money. They want to do this by charging you money for the privilege of having money.

Say what?

They want to charge you negative interest rates on the money in your bank account. Rather than your money in the bank growing every year because the bank is paying you interest, the bank will now charge you interest every year

---

[28] Martin Armstrong, "Negative Interest Rates & Eliminating Cash — The Summers' Solution", Armstrong Economics, November 17, 2013, http://armstrongeconomics.com/archives/15764.

and give the interest to the government. Your money in the bank will constantly decrease.

In other words, the government is going to sneak in extra tax on your savings without calling it a tax. And they will tax it and tax it and tax it every year it is in the bank.

Ha, you say. I'll just pull it all out and store it under my mattress as cash.

But First, Eliminate Cash

And so we return to our good friend and " benefactor" Larry Summers. He wants to eliminate cash. He says we should move to a completely cashless society where you *must* keep all of your money in a bank.

Of course, the government can have easy access to your bank account. It can take half[29] of your account if it wants (as they ended up doing in Cyprus). Or it can charge you negative interest rates.

> "Government will be able to tighten the vise on its people one more turn, restricting their freedom of choice (how to pay), wiping out any privacy in those transactions, and imposing another layer of government control." — Wolf Richter

In other words, the government is setting itself up to take complete control of your money.

You will have no recourse.

You will have no way to protect your earnings.

You will have no say in how much you get to keep.

And your freedoms? Yeah, forget those. They're gone with the greenback.

Of course, Big Government loves this idea.

Law enforcement thinks it's *wonderful* that all of your transactions *must* be disclosed to a third party. In other words, you will have NO privacy-ever again.

---

[29] "What has been agreed in Cyprus?" CNN, March 27, 2013,
http://www.cnn.com/2013/03/18/business/cyprus-bank-levy-explainer/

> "Physical paper money provides the check against negative interest rates for if they become too great, people will simply withdraw their funds and hoard cash. Furthermore, paper currency allows for bank runs. Eliminate paper currency and what you end up with is the elimination of the ability to demand to withdraw funds from a bank." – Martin Armstrong

Leftist, statist wonks gush out the wonderfulness[30] of a world where you have no protection against government and banking fraud. They're so happy to tell you how this will fix all of the problems with the economy that were caused by government misspending, government interference, and the banks making bad loans.

They don't stop to consider that this will lock the poor[31] out of the worldwide economy completely. Why? Because, of course, the poor are too poor to use banks. There is a huge population in this world that is "unbanked". That is, they have no access to banking services due to their poverty. In a cashless world, the poor are not only unbanked, they are completely unable to use or keep any money whatsoever.

The gung-ho cheerleaders for a cashless society don't seem to register the fact that there has never been a centrally-planned and centrally-controlled economy in the history of the world that has succeeded. Not one.

But the biggest problem here is the very mindset that even conceives of an idea like this. Mr. Summers, and those who agree with him, are of the viewpoint that all of your money belongs to the government and it can be seized whenever they want for whatever reason they might imagine they have.

In other words, you are serfs or slaves of the almighty state. You are not free citizens. You are subjects to be ruled not citizens to be represented. Therefore, whatever you have belongs rightly to those in power to use as they please. This is the underlying and unspoken assumption in Mr. Summers' idea. If Mr. Summers and his gang of thieves viewed you as free citizens who held and controlled private property, he wouldn't even begin to advance the idea that the government could charge you negative interest rates on your own money that you earned through your own hard work and investment. Mr. Summers *has* to see your money as belonging to the government and you as little more than a serf in order for his idea to even be conceivable.

---

[30] Matthew Yglesias, "Mo Money, Mo Problems: How eliminating paper money could end recessions", Slate, December 12, 2011, http://www.slate.com/articles/technology/technology/2011/12/how_eliminating_paper_money_could_end_re cessions_.html

[31] Conor Friedersoorf, "The Hubris of Trying to Eliminate Cash", The Atlantic, June 6, 2014, http://www.theatlantic.com/business/archive/2014/06/the-technocrats-who-want-to-take-your-cash-away/372322/

What Should Be: Limited Government

The people who created the United States of America were very wary about the whole idea of government. What they did not want was exactly the kind of government we have today. They designed a government that would play only a limited role in citizens' lives.

One of the very reasons that the United States because the richest nation in the history of the world is that it had a limited government. The Founders were so skeptical of the government that they created a Constitution that kept the government from interfering in the lives of its citizens. The Constitutional freedoms the Founders gave us are vital to free markets.

Simply put: free markets *require* freedom because free markets *are* freedom. A free market is just a large group of people using their money in a way that they think will benefit themselves the most. When people are free to choose for themselves, the majority of them will choose in a way that makes the entire society more prosperous. Constitutional restrictions on the government give us exactly that freedom. We have the prosperity we have precisely because the government has been so limited for so long.

Of course, the limited size and power of the government has been changing since the early 1900s. The more the government grows and the more it gains power, the more unstable our economy becomes.

What Can Be Done?

So what can we do to move from what is to what should be?

As you might expect, we first need to see the root cause of all of these problems. And as the old saying goes, to find what's really causing the problem we have to follow the money. So that's exactly what we'll do.

In this book, we'll propose real solutions based on a solid understanding of what the root problem really is.

And what *is* the root problem?

21

It's simple. Our money itself is broken. Yes. It's the *kind* of money we use that's driving all of this.

MONEY IS WHAT WE SAY IT IS

Most people think of money as being something like air. It's just there. It is what it is. But that's not true. There have been other kinds of money in the past that worked *better* than the money we have now.

And I'm not talking about the form that the money takes. In other words, money can be exactly the same thing whether it's represented by gold coins, slips of paper, glass beads, disk-shaped rocks (yes, disk-shaped rocks have been used as money), hard-to-obtain seashells, or digital bits in a computer. It's not the physical form that determines what money is. It's the features that the money has.

The really radical idea that this whole book is based on is that money is what we say it is. It has the features that we all agree it has. As we'll see in this book, we can use the blockchain to create better forms of money that solve problems rather than propagate them.

> "Money is a collective agreement. If enough people come to the same agreement, what they agree upon becomes secondary, whether it be farm animals, gold, diamonds, paper, or simply a code. History proves all these cases to be true. Who knows what the future is going suggest to us as money, once we see digital currencies as ordinary?"
> – S. E. Sever

As you read this book, you'll see that the kind of money we use concentrates wealth from the masses to the elites. It concentrates power from the people to the government. It creates artificial levels of competition in the economy, drives overconsumption, and kills the free market.

In fact, unless we fix the money we use, no other steps we take to solve the problems we've discussed in this chapter will make any difference at all. And the sad reality is that most of the measures we attempt will actually make our problems *worse*.

EXAMPLE: GOVERNMENT ANTI-POVERTY PROGRAMS

For instance, think about government anti-poverty programs. The US has had a declared "war on poverty" for fifty years now. In that time, this country has spent more than *four trillion dollars* fighting poverty. What has been the result?

Greater poverty, that's what. By any sensible measure, the war on poverty has been a complete and abject failure. Poverty has grown beyond anything we could have imagined when the war on poverty started.

> "The most experienced charity workers regard [public relief] as a source of demoralization both to the poor and the charitable. No public agency can supply the devoted, friendly, and intensely personal relation so necessary in charity. It can supply the gift, but it cannot supply the giver, for the giver is a compulsory tax rate." – Mary Richmond

Simply put, if government anti-poverty programs could possibly work, they would have worked by now. Half a century of government programs aimed at eradicating poverty have only caused a massive increase in poverty. Literally tens of millions of people in the US are now trapped in a grinding poverty from which they cannot escape. It is in fact inhumane to foist more government programs on people and keep them caged into a lifestyle that destroys any hope they have for their future and drains away their dreams of a better life.

While it may sound good to give handouts to the poor, we have to remember that doing so cuts them off from the possibility of being successful. It traps them in poverty because they can't sustain the earning power to maintain the lifestyle that the handouts give them. By simply giving the poor handouts and not providing them with a real path to prosperity, we are trapping them in a life of grinding poverty from which they cannot escape. This is an unspeakably inhumane cruelty that we are perpetrating upon our fellow citizens in an effort to make *ourselves* feel good while doing nothing substantial. We can falsely believe that we are helping and look at ourselves as being moral when in fact we are destroying people's lives and creating a monstrous injustice.

> "Where justice is denied, where poverty is enforced, where ignorance prevails, and where any one class is made to feel that society is an organized conspiracy to oppress, rob and degrade them, neither persons nor property will be safe." - Frederick Douglass

With the blockchain, it really isn't that hard to provide the poor with a viable path out of poverty. Failing to do this for our friends and neighbors is completely unjust.

It is literally *insane* to keep trying to solve the problem with government programs when government programs have proven themselves to not work. It's time to think "outside the box" as the cliché goes. We have to look at the problem in a completely different way than we ever have before.

If we fix our money, the poor can have a real, attainable path out of poverty. It won't be a cakewalk for them. Nothing worthwhile in this life is easy. Anything that gets handed to us on a silver platter is either worthless or comes with strings attached.

> "I am for doing good to the poor, but I differ in opinion about the means. I think the best way of doing good to the poor is not making them easy in poverty, but leading or driving them out of it." - Benjamin Franklin

But if we do use the blockchain to reinvent the kind of money we use, we can give *everyone* in our economy a fair chance to rise from poverty. We can even create tools that will give them a boost along the way. And we can do this without government programs or regulations, without taxes, without going into debt, and without even asking anyone for charity. There are real answers out there that have proven themselves to work if we will only learn the lessons of past experience and apply new technologies. History has shown that massive social problems can be solved. But we first have to dig deep into the real source of these problems. T

REAL PROBLEMS, REAL SOLUTIONS

The problems presented so far are very real and very debilitating for a large percentage of the human population. All over the world, people are increasingly forced into grinding poverty and near slavery (or sometimes outright slavery) when they could have prosperity and liberty.

As we go forward in this book, we'll first examine our money and banking systems and see how and why they generate many of the problems we face in our society. We'll see why new economies, based on decentralized, distributed, self-organizing blockchain systems are more resilient, more sustainable, fairer, and more humane than what we have now. In addition, we'll look at why limited government and the free market will solve a lot of the problems we face.

This book will also discuss the primary alternative to limited government and the free market, which is Marxism. Marxism includes socialism and communism. We'll see that these defeatist philosophies can't possibly work, and in fact are just a relabeling of the system we have now. They not only won't change anything for the better, they'll make things worse.

One of the most startling assertions this book makes is that free markets died a long time ago. Since then, we've mixed an increasing amount of Marxism into our economy. And that's precisely what's driving most of our problems.

However, there are real solutions. The blockchain enables us to democratize the economy. It lets anyone apply their brainpower to the problems we all face. They do so with the idea of making money, but so what? Because the blockchain enables small nonprofits, small businesses, and even individuals to easily deploy worldwide solutions, it is possible to crowdsource large-scale problems such as poverty. Literally millions of people around the world can deploy possible solutions. And it only takes a few that succeed to make those problems go away completely.

Do you remember when no one really knew what the Internet was? Now we use it in almost everything we do. The blockchain is the same way. As of this writing, the blockchain is about where the Internet was in 1987 or 1988. Only a few very technical people know about it. But companies, individuals, and large financial institutions clearly see the potential of the blockchain. They are already investing billions of dollars in blockchain projects.

> "Any strategy to reduce intergenerational poverty has to be centered on work, not welfare--not only because work provides independence and income but also because work provides order, structure, dignity, and opportunities for growth in people's lives." - Barack Obama

Why? Because the blockchain is one of those rare technologies that completely changes everything. One of the many implications of the blockchain is that we can use it to apply free market principles to money itself. If we apply free market principles to money itself, We The People can reinvent our money system to serve us rather than the elites. Right now, we live in an economic monopoly. But we can restore economic democracy with new forms of money.

Now before we go on, let's make one thing clear. The free market can't fix all of our problems on its own. But by reinventing our money system, we can provide ourselves with tools that will enable us to directly attack daunting social problems and resolve them without limiting anyone's freedom, without a forced redistribution of wealth, and without growing the government. In short, it's a free market money system that promotes liberty and liberates us from all forms compulsion.

We can provide ourselves with new forms of money and with economic tools that will promote freedom and liberty.

> Innovation distinguishes between a leader and a follower. --Steve Jobs

We can create new currencies and new economic systems that will provide a real path out of poverty for anyone who will work.

We can open up new opportunities for investment that can be scaled down to the budget of the smallest investor. Everyone in our economy can have the kind of income-producing assets that are currently available only to big banks and Wall Street investors. Basically, we can bring the power of Wall Street to Main Street.

We can build specialized currencies that will solve massive social problems and make people more prosperous in the process. It's proven. It's actually been done before.

> "The American free market system is the greatest engine for prosperity and opportunity that the world has ever seen. Freedom works." - Ted Cruz

We can invent microeconomies that multiply the effects of the money we spend on such things as education. For instance, we can make $10 million spent on education do the work of $100 million. As amazing as it sounds, it's perfectly attainable with very little added overhead to the educational system (one employee and one computer per school).

We can make green energy more profitable and accessible to most of our civilization. It's actually possible, for example, for someone who lives in a sunny spot to become a small-scale green energy producer and make a tidy profit at it. But it will take a radical new form of economic infrastructure to do it.

As "pie in the sky" as all of these statements sound, they really are attainable if we are just willing to rethink what money is and if we are willing to apply the blockchain in ways that can quite possibly enable us to completely reinvent our entire civilization. And that is exactly what this book is about—applying the blockchain to seemingly insurmountable problems to create a better world. So let's begin to do just that by taking an honest look at how our money and banking systems affect us.

2 THE PROBLEM: OUR MONEY IS BROKEN

We are currently in a financial crisis unlike any the world has seen. Many nations across the world are approaching bankruptcy. As I stated in Chapter 1, the entire planet seems to be running out of money.

Many attribute the world's current financial problems to government mismanagement, not enough capitalism, not enough socialism, not enough communism, global banking conspiracies, and so forth.

While problems like government mismanagement and the debilitating effects of dysfunctional Marxist philosophies (socialism and communism) are real, they are actually symptoms and not the root cause of our current economic problems. Until the root cause is addressed, nothing else will make a lasting difference. And in fact, most solutions to our current problems, such as increasing government regulations, limiting personal liberties, and forcing various forms of Marxism on entire nations, only serve to make our problems much, much worse.

Our underlying problems are the banking system that we use and the kind of money we have.

PROBLEM 1: FRACTIONAL RESERVE BANKING

Our banking system was invented in the 1600s, when the science of economics was completely unknown.

By way of comparison, the leading medical authorities of that time recommended that doctors treat diseases by bloodletting–the process of cutting and bleeding a patient to release bad "humors" that were supposedly trapped in their bodies.

Likewise the disease malaria was thought to be caused by inhaling bad air. That's where the name comes from; mal-air-ia, bad air.

An era with such limited scientific knowledge could not possibly invent a banking system that would prevent the kinds of problems we see today. Instead, bankers–who were mostly aristocrats–invented a banking system that would enrich the elites rather than ensure the democratic principles that we value today. The system that they came up with, called *fractional reserve banking*, has served its purpose well for over 400 years. Even today, fractional reserve banking enriches the elites at the expense of everyone else.

> "It is well enough that people of the nation do not understand our banking and monetary system, for if they did, I believe there would be a revolution before tomorrow morning." – Henry Ford, founder of Ford Motor Company.

Most people, even those in the money-related professions such as accounting and finance, do not know some of the basics about how our economic system works. The problems created by fractional reserve banking are not very widely understood.

HOW OUR MONEY GETS CREATED

It is common for people to think that the nation's central bank, the Federal Reserve Bank (also called "the Fed") creates our money. While that is not incorrect, it is not a complete picture.

> "If the American people ever allow private banks to control the issue of their currency, first by inflation, then by deflation, the banks...will deprive the people of all property until their children wake-up homeless on the continent their fathers conquered.... The issuing power should be taken from the banks and restored to the people, to whom it properly belongs." – Thomas Jefferson

Actually, the Fed only creates a small percentage of our money. Banks create most of our money. And they create money through debt.

Suppose that I have somehow received $10 million from the Fed. In reality, this is not possible. But for the sake of this example, let's pretend that this could happen.

I take my $10 million and deposit it in a bank. What does the bank do with it?

Figure 2.1 illustrates the answer to this question. It shows that the bank holds 10% of the initial deposit and lends 90%, which amounts to $9 million[32]. The bank keeps that small fraction (10%) on reserve in case we want to withdraw

---

[32] This too is a simplified picture of how fractional reserve banking operates. The percent of the initial deposit kept on reserve by the Fed is called the *reserve requirement* or the *liquidity ratio*. Deposits of less than $12.4 million have no reserve requirement–the bank can lend it *all* out. From $12.4 million to 79.5 million, the reserve requirement is only 3%. Anything over 79.5 million is kept at a 10% reserve. However, I have used the 10% figure in this example for clarity and simplicity.

some of our money. That's what the "fractional reserve" part of the banking system name means.

What happens to the 90% of the money that the bank lends out? When someone borrows money, they inevitably deposit it in a bank. Or they spend it somewhere and the business that receives it puts it into a bank. So the bank[33] receiving the deposit of $9 million does the same thing. That is, it holds 10% and lends 90%. The amount of the new loan is $8.1 million.

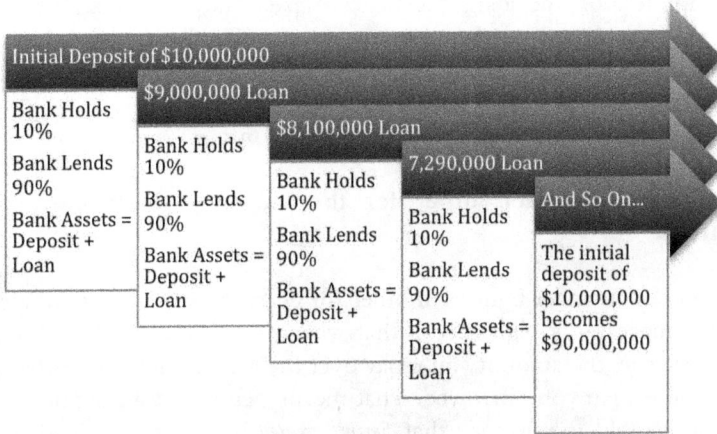

Figure 2.1 How Banks Create Money

Of course, the money continues to ripple through our banking system. Each time someone deposits it in a bank, that bank keeps 10% and lends 90%. In this way, the initial $10 million becomes $90 million.

WHAT ABOUT THE INTEREST?

Everyone knows that the bank wants to be paid back more than they lent out. For example, if you borrow money for a mortgage, you'll probably end up paying 2.5 to 3 times what you borrowed[34]. The extra money is interest and it is the vast majority of what you pay back.

---

[33] In this example, I'm more or less treating the entire banking system as "the bank" for clarity. In reality, it would be many banks but the effect is the same.
[34] In my grandfather's time, the amount you paid back on a mortgage was about 1.5-2 times what you borrowed.

Here's a question few consider: Who creates the currency for the interest?

In our example, the Fed created the initial $10 million for the deposit. The banks created an additional $90 million to lend out. Who creates the money that we need to pay the interest?

Answer: **No one**.

As strange as it seems, it's true. No one creates the actual currency needed to pay the interest on the loans the banks make. And remember, *most* of the money that people pay on their loans is interest. But the cash needed to pay back that interest was never created by anyone.

So how do we obtain money to pay the interest on our loans?

Answer: **We fight each other for the currency that's currently in circulation.**

That's right. You and I fight each other for currency at levels of competition that are unnecessarily high. Recall that when you take out a loan such as a home mortgage, the amount you repay over the life of the loan is often 2.5 to 3 times more than you borrowed. That means that the vast majority of what you repay is paid in currency that *wasn't created* by our money and banking system.

The fractional reserve banking system creates artificially high levels of competition with our economy as we fight each other for the limited quantities of currency that exist. If we had a functional money and banking system, it would be easier to become more prosperous. Fractional reserve

> "A man in debt is so far a slave." - Ralph Waldo Emerson

banking, since it was designed to enrich the elites, does not work well for the vast majority of people who have to live with it. It actually makes their lives harder than they should be.

CONSEQUENCES OF FRACTIONAL RESERVE BANKING

What are the natural consequences of our banking system?

Increasing Poverty

Fractional reserve banking forces artificial levels of competition on anyone who participates in the economy. Because the currency to pay the interest is not produced by anyone in the system, there is not enough currency to pay back all of the loans being made.

If you were to count up all of the physical currency issued by all countries that is in circulation right now, it would total up to about about $4-$6 trillion when expressed in US dollars. The current US national debt is $19 trillion and is rapidly approaching $20 trillion. The rest of the currency is electronic bits floating around our computer systems that was created through debt. In other words, there is not enough real currency to pay back all of the current debt of just the United States of America, let alone the rest of the world.

So we all fight it out for what currency we can get. Under fractional reserve banking, some people *will* fall into poverty no matter how well they manage their financial affairs. In order for the

> "In a country well governed, poverty is something to be ashamed of. In a country badly governed, wealth is something to be ashamed of." - Confucius

system to function, someone *must* lose everything. Nothing else will keep it going. And because the bankers receive more money than they lent, they gain control over an increasing percent of the available currency. This in turn decreases the amount of currency that can be used to pay back loans. As a result, competition increases and so does poverty. Over time, the gap between the rich and the poor increases and the upward mobility out of poverty decreases. This is a fundamental inequity in our current system that can't be fixed no matter how many taxes we levy or how many laws we change or how many handouts we give. It is downright cruel and it is completely unnecessary.

And the people who are most likely to fall into poverty under the fractional reserve banking system are the most economically vulnerable. This generally includes the

> "Poverty is the worst form of violence. Mahatma Gandhi

elderly, divorced/single women with children, and ethnic minorities. For example, people of African descent have long been at a significant economic disadvantage because of hundreds of years of racism. But although these groups of people are impacted first by the negative effects of our current system, over time those effects spread to nearly everyone. And they will certainly spread to everyone but the highest financial elites.

As poverty increases, social mobility–the ability of people to move up in the world through hard work, innovation, and education–dries up. People

31

become hopeless and often act out violently. Fractional reserve banking will eventually cause so much poverty and so completely destroy upward social mobility that the oppressed underclass will resort to violence against the so-called 1% in an effort to end the grinding misery of their lives.

Concentration of Wealth from the Many to the Few

Fractional reserve banking concentrates money from the many to the few because all money is *borrowed* into existence. This is true even for money created directly by Fed. All of our money is generated by debt. The biggest banks borrow it directly from the Fed. They, in turn, lend to smaller banks. These banks do the same and the money gets repeatedly lent out until it reaches us.

At every level of this multi-tiered system, the banks doing the lending charge interest (so does the Fed itself). *All* of that interest gets paid back by the person or business at the bottom who borrows money to buy a house or invest in a business. That is you.

You pay back *all* of the interest generated through layer after layer of borrowing. And the money you pay flows upward through this pyramidal lending structure until it reaches the lenders at the top.

"Our whole system of banks is a violation of every honest principle of banks. There is no honest bank but a bank of deposit. A bank that issues paper at interest is a pickpocket or a robber. But the delusion will have its course. ... An aristocracy is growing out of them that will be as fatal as the feudal barons if unchecked in time." – John Adams

Keep in mind that the lenders at the top just create the initial money out of thin air. We're not on the gold standard any more so they're just lending out money with no backing or value to it. And they want to be paid back by money that we back with our labor and industry. In other words, the money they lend to us is valueless and we contribute value to the economy to obtain money. When we do, the money that we hold essentially becomes backed by the value we created. We then use that to pay back the original debts. The elites essentially use money as a tool to concentrate value that you and I create to themselves. The only possible outcome of that is the concentration of wealth from the many to the few.

We are starting to see a skyrocketing income gap between the rich and the poor[35] that is a direct result of fractional reserve banking. This was not noticeable when our current system was first introduced. However, after 100

---

[35] See census.gov: Table IE-6. Measures of Household Income Inequality: 1967 to 2001 (http://web.archive.org/web/20070208142023/http://www.census.gov/hhes/www/income/histinc/ie6.html).

years of fractional reserve banking, the cumulative effects of the system are plainly apparent[36]. Figure 2.2 illustrates the huge income gaps that our fractional reserve banking system has generated in recent years. The top 1% of income earners has nearly tripled their income since 1979. Yet income levels have largely been stagnant for the rest of us.

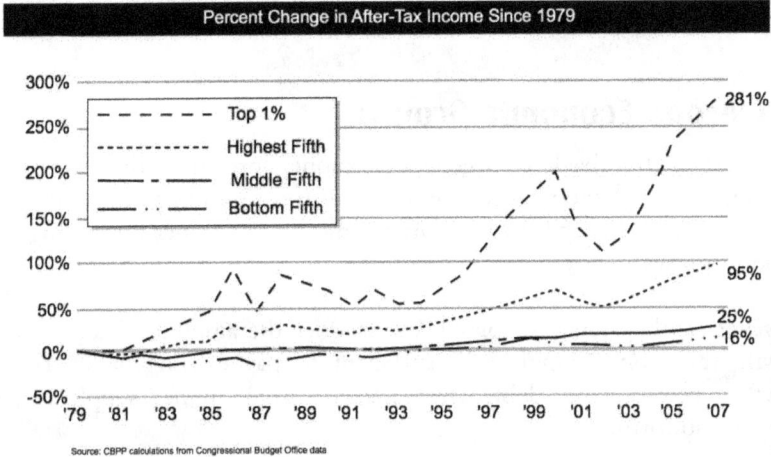

Percent Change in After-Tax Income Since 1979

Source: CBPP calculations from Congressional Budget Office data

Figure 2.2 Income has stagnated for everyone but the top 1%.

In a healthy free market economy, there is *always* a difference between the earning power of the top earners and the bottom earners. However, this is exactly what spurs the bottom earners on to greater earning power. If everyone has a fair chance at moving up to a higher income level, then everyone who has the drive and initiative can do so. The lazy get left behind, but that is exactly what they choose by being lazy. Everyone else can move forward and move up in the world.

However, when the gap between the top and the bottom becomes too large, it indicates something is wrong. This should not occur in a healthy economy. If wealth is so concentrated into the hands of a small percentage of the population that real upward mobility is impossible, then the economy has simply become unsustainable. It will eventually collapse.

---

[36] Of course, fractional reserve banking is not the *only* cause of this inequity, as any economist worth his or her salt will tell you. But it is certainly one of the *primary* causes. And it is also one of the most unrelenting sources of income inequality in our lives.

A healthy economy provides a path out of poverty for those who will work hard, be thrifty, and gain more education. But increasingly, the world is losing site of this path to prosperity because too much wealth has concentrated into the hands of too few people.

Simply put, fractional reserve banking was designed to enrich the elites over time. It does that very efficiently—at a terrible cost to those of us who are *not* in the top 1%.

Cancerous Economic Growth

The need to pay back debt-generated money pressures the economy to constantly grow. A growing economy creates more money so the competition for circulating currency is less[37]. More people get to prosper for as long as the economy grows.

However, as the money supply grows, so does the amount of debt. Without growth, the money supply isn't sufficient to pay existing debts. But that growth generates more debt, which needs a larger money supply to pay it back, and so forth.

On the positive side, economic growth does bring a higher standard of living, advances in technology, and so forth. However, unending growth also consumes increasing amounts of resources. Over time, the resources become scarce in relation to the demand. National and international conflicts quite naturally increase.

The unrelenting need for growth means that businesses must constantly sell more. They need to get consumers to buy more. They do this through a variety of methods, but the result is that we consume far more than we need to in order to maintain our standard of living. Our banking system drives overconsumption. Our economy behaves essentially the way cancer behaves.

"Growth for the sake of growth is the ideology of the cancer cell." – Edward Abbey

In a healthy economic system, we can achieve a steady state. Or more accurately, we can have an economy that matches our needs. So it can grow if the population grows, shrink when the population shrinks, and maintain a steady state when the population does.

---

[37] This is a generalization. Competition is only less when the money is initially released because new money also means new debt. The weight of that new debt eventually catches up with everyone in the economy.

A healthy economic system does not require unrelenting growth. Nor does it inhibit a rising standard of living. It is actually possible to have a rising standard of living without a growing economy. That doesn't make sense to those of us who've been raised for generations with the philosophy of "growthism" in our economy. Nevertheless it's true.

The reason we can have a rising standard of living without the cancerous economic growth we are forced to endure now is simply because over time prices go down. Improvements in farming methods decreased the price of food. Improvements in the production of silver lowered the price of silver. Improvements in the production of computers has caused the price of computers to go down. Everything in the economy works this way.

Another way to say this is that growth in technology and scientific knowledge enable us to have a rising standard of living without forcing the economy to grow. As technology advances, it's actually possible for us to consume less while having a better standard of living. Unrelenting economic growth is not necessary.

> "What consumerism really is, at its worst is getting people to buy things that don't actually improve their lives." - Jeff Bezos

Make no mistake, there are times when the economy *should* grow. For instance, when the population rises, the economy needs to grow to match it or people will find themselves in poverty. But in a healthy economy, growth is not *always* essential for a rising standard of living.

Ecological Devastation

At some point, we must reach the limits of the Earth's capacity for the economy to grow. What will we do then? Any slowdown in the growth of the economy adversely affects all of its participants because without growth there isn't enough money to pay existing debts.

If we were to hit a limit to growth, our economy would have to attain a steady state. But because of fractional reserve banking, it can't be in a steady state and continue to function. A stop in the unrelenting growth could throw us into economic chaos. The consequences of an economic contraction could be even worse.

Some people think that they can fix the ecological problems by simply limiting economic activity. Most of their efforts are so harmful to the economy that they endanger the stability of civilization.

For example, some think that they can drive people to green energy by making fossil fuels so expensive that green energy actually becomes a cheaper alternative. This is a dangerously naïve approach to helping the biosphere. The only possible result of such a policy is economic collapse. And when the economy collapses, everyone pours out into the countryside to secure for themselves whatever resources nature can provide. In other words, the wilderness is pillaged in a way that's so devastating that it may never recover.

Simply put, economic disasters quickly become ecological disasters. But in a healthy economy, we can achieve a greater balance between what we want, what we need, and what we have here on planet Earth.

Economic Diversity Dies

I'll repeat here that all of the money in the fractional reserve banking system is created by debt. This is even true of the money issued by the Fed. It lends that money out to big banks. Those banks lend it to smaller banks, and so on until it gets down to you and me.

Debt-generated money gives advantages to big businesses. The growth it forces upon the economy means that businesses must become increasingly efficient over time. Efficiency is normally a good thing. However, the drive for growth forces levels of efficiency upon companies that can often only be achieved by large businesses. For example, local businesses often cannot compete with large, big-box retailers such as Walmart and Target. Over time, larger businesses decrease economic diversity in ways that have profoundly negative effects on local economies[38].

> "Behind every small business, there's a story worth knowing. All the corner shops in our towns and cities, the restaurants, cleaners, gyms, hair salons, hardware stores - these didn't come out of nowhere." - Paul Ryan

It is also possible to become so efficient that a business loses its resiliency to deal with changing market conditions. Woolworth's, Borders, and Digital Computer Corporation (DEC) are just a few examples. These were all well-run and efficient businesses providing good products and services. But they were so focused on one way of doing business–the way that had proved most efficient until that time–that they lost the ability to deal with changes in the marketplace. They had no resiliency.

The cumulative effect of this is that small business, which has traditionally been the main engine of our economy, is slowly dying while big business, which actually employs fewer people, takes over. Because there are fewer

---

[38] See investopedia.com: Wal-Mart Effect (http://www.investopedia.com/terms/w/walmart-effect.asp).

businesses supplying any given product, the diversity of products available decreases. As the number of businesses decreases, it becomes increasingly likely that the demise of a few big businesses could have serious negative impacts on the entire economy. We see this very clearly in the banking sector with the idea of "too big to fail" banks that the government insists must be bailed out by taxpayers when they hit the verge of collapse. It does not matter that their imminent demise is a direct result of their own mismanagement or simply failing to adapt to changing market conditions.

In a functional free market, economic diversity increases. The economy is far more resilient because there are more employers making a greater diversity of products that have the possibility of succeeding. But debt-based money and fractional reserve banking make this nearly impossible to achieve in the long term.

A POSITION OF PRIVILEGE

Banks have a unique and privileged position in our economy. When new money is released, they get it first. They also get it more cheaply than any other type of business. As such, it enables them to expand their business before anyone else gets a chance. It gives them a jump on every other industry in our economy.

Because of laws and government regulations, banks have very little competition. Think about it. Almost the only transaction processors in our economy are the banks and the credit card companies. That's pretty much it[39].

> "I believe that banking institutions are more dangerous to our liberties than standing armies." — Thomas Jefferson

Ever wonder why credit card transaction costs are so high for merchants? Now you know; it's lack of competition. Ever wonder why banks can get away with charging so many fees and even demanding a copy of your tax return when you get a mortgage? Yup. It's because they have no real competition. They don't have to care. You've got nowhere else to go.

The lack of any other alternatives to the fractional reserve banking system means that bankers can use their privileged position to concentrate wealth to themselves. Figure 2.3 summarizes how fractional reserve banking works.

---

[39] The rise of online purchasing is creating greater competition for banks and credit card companies. These days, retailers such as Amazon are creating their own payment systems. But they still interoperate with the banks and credit card companies, as do other payment processors such as PayPal.

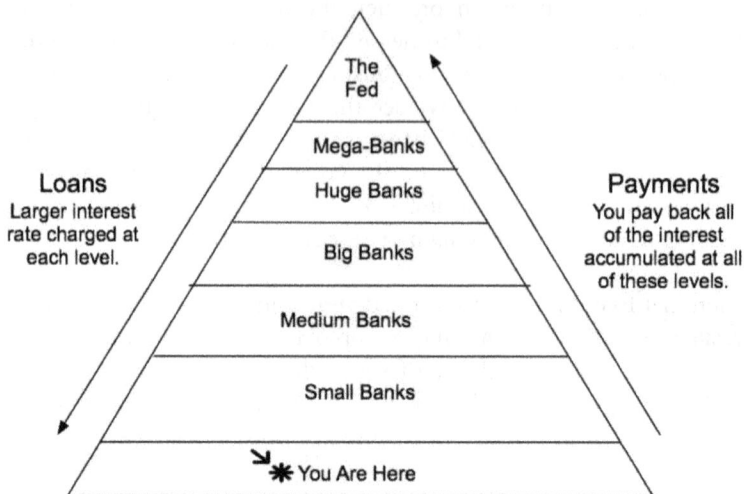

Figure 2.3 The Fed lends money into circulation and every level of lending in the system requires that more money be paid back than was lent out. Because you are at the bottom of the system, you have to pay *all* of the interest on *all* of the loans.

Looking at Figure 2.3, it's easy to see how banks use their position of privilege in the economy to concentrate wealth to themselves. As the arrow on the left of the diagram shows, loans from the Fed propagate downward to you and I. But payments concentrate upward to the bankers. Far more money goes up to the bankers than came down to you and I. In a system like this, you can clearly see why poverty grows over time.

And its even worse than it appears.

It used to be the case that the money issued by the Fed was backed by gold. In other words, the money represented the value of the gold. But these days, the money that moves down the pyramid in Figure 2.3 is backed by nothing. Another way of saying that is that **the money has no value at all**.

The Fed is playing a slight of hand with you and I. New money issued by the Fed **has absolutely no value at the time of its issue.** The Fed's valueless money then gets lent through the banking system to you and I. When the money arrives at the bottom of the pyramid, you and I have to work in order to pay it back. That means that the money we pay back the loans with–the money that goes back up the pyramid–is backed by the labor of you and I.

Just to be clear: the money that comes from the Fed is backed by nothing. It represents no value at all. But when you and I put our labor, our effort, our innovation, and our work into the economy in the form of goods or services, the money we receive then represents our effort, innovation, and work.

The Federal Reserve Bank does not have any value to put into its money, you and I do. Money only represents the goods and services it can buy. The only way people can buy goods and services from the economy is if they put goods and services into the economy. The Fed doesn't do that. It actually adds nothing to the economy but paper with fancy stuff printed on the front and back[40]. Only you and I can put value into the economy through the goods and services that we sell to our employers or our customers. The Fed can't do that.

But here's the really evil thing about the Fed. It lends us valueless pieces of paper. We associate value with that paper by putting our work into obtaining enough of it to pay back our loans. We then send that paper back up the wealth pyramid. It now carries value. That value is transferred back up to the elites at the top of the pyramid.

It's important to understand that money and wealth are two different things. Money can represent wealth. But the only thing that produces wealth is selling goods or services. The Fed never produces wealth. It only produces money. The money it produces never has any value when the Fed releases it. It acquires value when we sell goods and services into the economy in order to obtain it. When we do that, the money we have then represents real wealth. It represents the goods or services we put into the economy. Then that money goes back up the pyramid, carrying our wealth with it.

Over the long term, the only possible outcome of this economic system is the impoverishment of people like you and I and the enrichment of the elites at the top.

> "Money plays the largest part in determining the course of history." - Karl Marx.

Once you realize how our system of fractional reserve banking really works, it's easy to see the underlying systemic causes of our growing poverty. It's also easy to see that it's all a scam.

And it gets worse. You and I can only create so much value at the bottom of the pyramid. But the Fed can dump all the paper it wants on us. Any time it does, the value of each piece of paper decreases because there are more pieces of paper representing a limited pool of wealth. So the Fed and its associated

---

[40] You may rightly object that most money is not paper or coins, just bits in computers. You're correct. But I'm going to pretend it's all paper and limit this discussion to only paper money for clarity.

banks can not only suck away our wealth through fractional reserve banking, it can also drain us dry by inflating our money. And that's what we'll talk about next.

PROBLEM 2: GOVERNMENT CURRENCY MONOPOLY

The second problem that is driving most of the world's current difficulties is that the governments of the world have granted themselves a monopoly over the production, distribution, and regulation of our currency. They have obliterated economic democracy in favor of an economic monopoly that favors those in power. This monopoly gives a government total control over the value of the nation's money. The government then decides what its monetary policy will be.

It's important at this point to stop and distinguish between *monetary policy* and *fiscal policy*. Monetary policy decides how much money the government will print. Fiscal policy decides how to spend the money once it's printed.

WHAT IS ECONOMIC DEMOCRACY?

Currently, many nations in the world are political democracies. However, at this time, there are few if any economic democracies. Nearly every nation uses one dedicated currency that is controlled by a governmental monopoly. For this reason, everyone in the world is tied to the fate of the currency of the country in which they reside. Governmental policy can wipe out a citizen's life savings in a matter of hours. **Control over currency is control over all aspects of human existence.**

In a true democracy, both political and economic freedoms are available to all citizens. Buyers and sellers are free to choose to use whatever currencies they prefer for their transactions. Modern technology makes multicurrency economies easy and straightforward for everyone involved.

And here we are, back to talking about the blockchain again. The blockchain is a technology that can be used to create a multicurrency free market–an economic democracy. And that's just the beginning of what you can do with it.

HOW DID WE END UP WITH AN ECONOMIC MONOPOLY?

When the USA was created, it was an economic democracy as well as a political democracy. We have abandoned economic democracy in favor of an economic monopoly. Let's take a look at how that happened.

It is Illegal for the US Government to Issue a Currency

Did you know that the government of the United States of America cannot legally issue a currency? It's true. The Constitution specifically prevents it. E. C. Riegle, an economist of the early 20th Century whose writing on inflation and government currency monopolies were nearly prophetic in nature, explains it this way.

> All of the thirteen American Colonies legalized the issuance of "money" by government, and all thirteen units of account passed out into thin air through total inflation–the inevitable result when [inflation] is carried to extremes. Following these Colonial experiments came that of the Continental Congress, from which sprang the continental, object of the reproachful phrase, "not worth a continental."
>
> It is not surprising that, with these horrible examples … before them, the delegates to the Constitutional Convention resolved to withhold from the federal Government this perverting power. The question arose when Article 1, Section 8, Paragraph 5 was up for discussion. This provision, as adopted, reads:
>
> > Congress shall have the power to coin money, regulate the value thereof and of foreign coin, and fix the standard of weights and measures.
>
> The clause as first presented included the words, "emit bills of credit." After debate, the delegates voted to strike out these words, and thus the Government was denied the power to issue currency. In those days currency was called bills of credit…
>
> The clause, as enacted in the Constitution, authorized the Government to "coin money," but not to issue it. It meant that the Government was empowered to set up a mint to stamp out coins from metal brought to it by private owners. The coins minted were not Government property; they remained the property of the citizen from whose metal they had been coined. He was thereafter entitled to issue these coins into circulation bearing the Government's guarantee of weight and fineness. To "regulate the value thereof" meant to define what constituted a dollar and its fractions. It did not mean to regulate the power thereof, as this would involve price fixing, an impossible task. (Flight from Inflation, p. 41: E. C. Riegle, The Heather Foundation, Los Angeles, CA)

In other words, the framers of the Constitution had already seen both the individual states and the Continental Congress issue currencies and inflate them into worthlessness. So they wanted to prevent the constitutional government from doing the same thing. To accomplish that, they put the US mints under the Department of Weights and Measures. This was so that the Department of Weights and Measures could determine how much gold or silver should be in a $1 coin. That was the meaning of "to regulate the value thereof" in the Constitution. The government's place was to figure out how much gold or silver was required for a $1 coin and certify that the coins produced by the mint actually had that much gold or silver in them.

Anyone who owned gold or silver could bring it to the mint. It would then be refined and pressed into $1 coins. The coins were not owned or issued by the government. The owner of the gold or silver was the owner of the coins. The seals and insignias on the coins simply verified that the coins contained $1 worth of gold or silver. That was the only involvement the constitutional government could have in coinage.

For the first 70 years of US constitutional government, that pattern was followed. Private banks were freely able to issue their own currencies. These currencies competed on an open market. Because of the difficulties in that era of time, distance, and the burden of handling many different types of paper and metal currencies, the free currency market was much more limited than

> "The Constitution is the guide which I never will abandon." - George Washington

the digital markets that are possible now. However, for that entire time, government-issued currency was seen as neither necessary nor desirable.

The Reinterpretation of the Constitution

If the Constitution prevents the US government from issuing a currency, what happened to change that?

The answer is the Civil War. Riegle explains it thus.

> The Civil War emergency, however, induced Secretary of the Treasury Salmon P. Chase to recommend to Congress the issuance of United States notes, popularly called "greenbacks," and Congress obliged. This was the first [money] issued by the United States Government, and it was frankly recognized as unconstitutional. It was justified on the ground of national emergency by Chase, although later, as Chief Justice of the Supreme Court, he condemned it in a majority report as unconstitutional. By a still later decision, however, with Chase this time in dissent, the Court sanctioned the practice and thus read into the Constitution what the founders had deliberately voted to keep out. (Flight from Inflation, p. 41: E. C. Riegle, The Heather Foundation, Los Angeles, CA)

By reinterpreting the Constitution—essentially treating it as a "living document" rather than a set of guiding principles that preserve the rule of law—the political wonks of the late 19th and early 20th Centuries made it seem like the Constitution said the exact opposite of what it actually did say.

It is most notable that the person who wanted to issue a currency to see the nation through the emergency of the Civil War was adamantly opposed to allowing government the same power in peacetime. He knew as well as the Founding Fathers that such a power corrupts the government.

> "Money is only a tool. It will take you wherever you wish, but it will not replace you as the driver." - Ayn Rand

For at least the last 2,000 years, there never has been a currency controlled by a government monopoly that has retained its value over time. Every single currency of this type has ended in hyperinflation that then caused a complete economic collapse. That collapse was typically followed by either civil or international war. Often it was both. A governmental monopoly on currency will always end in this exact same way.

The Advent of the Federal Reserve and the Abolishment of Private Currencies

To facilitate the creation of a national currency, the US government also created a central banking system called the Federal Reserve Bank on December 23, 1913. Over time, the Fed, has expanded its powers over nearly all aspects of economic life.

Private currencies issued by banks are called *bank notes*. Bank notes used to be common in the US and many other countries. The last bank notes in the US were withdrawn from circulation in 1957. It's interesting to ask why bank notes are illegal today. If they caused no damage to the economy from the adoption of the US Constitution to 1957 (168 years!) then why were they outlawed?

> "The modern banking system manufactures money out of nothing. The process is, perhaps, the most astounding piece of sleight of hand that was ever invented. Banks can in fact inflate, mint and un-mint the modern ledger-entry currency." - Major L L B Angus.

The answer of course is control. The creation of the Fed provided the government with the ability to consolidate its power over the economy in a way that was never possible before. But this can only be accomplished if there is no alternative currency for the citizenry. Therefore, it was necessary to have a complete governmental monopoly over currency in the United States (and all other countries, for that matter) to attain the degree of power the Fed offered. If the citizens of a country have access to a currency that does not inflate, they will cease to use a national currency that does. The reason for this is simple; they do not want the value of their paychecks and their life's savings to be diminished by inflation if they can avoid it.

43

The citizenry of every country with a central bank should be very worried. Under the management of a central bank, the buying power of the nation's currency *always* decreases. This is true of every nation on earth; the US is no exception. Under the management of the Federal Reserve Bank, the buying power of US dollar has decreased by an astounding 98% in the last 100 years, as shown in Figure 2.4.

The loss of 98% of the buying power for the US dollar[41] cannot be seen as anything other than complete and abject failure on the part of the Federal Reserve Bank and our current economic system. No one should be naive enough to think that the Fed can be trusted to manage the value of the dollar at all.

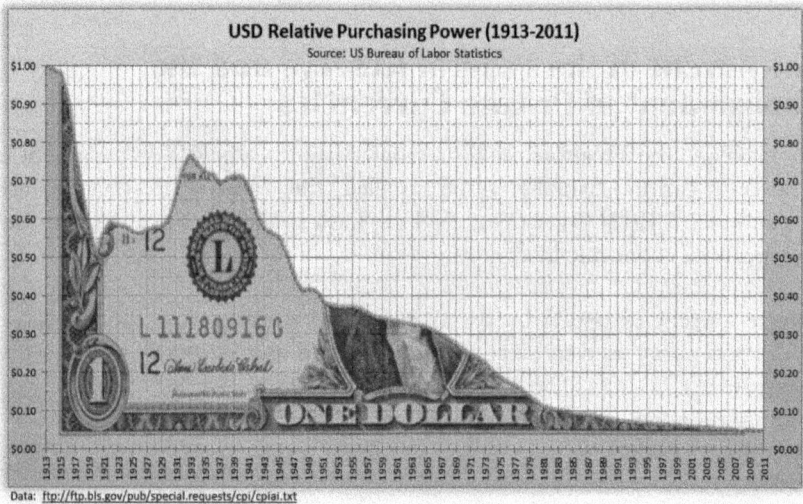

Figure 2.4 The buying power of the US dollar has decreased by 98% in the 100 years that it's been managed by the Federal Reserve Bank.

It would also be naïve to think that the dollar will ever recover from this precipitous loss in the buying power. That ship has sailed.

CURRENCY CONTROL AND MONETARY POLICY

The idea of "legal tender" is a complete fiction. It somehow implies that only government money is safe to use. This is merely government propaganda.

---

[41]For an interactive inflation calculator, see http://www.usinflationcalculator.com/.

Governments pass legal tender laws because they need a monopoly on currency to implement their monetary policies. They want the ability to create monetary policy for specific reasons. First, governments use monetary policy to play Santa Claus. Second, governments use monetary policy to temporarily suppress unemployment. Third, governments use monetary policy to try and control trade imbalances. Let's look at each of these.

> "Banking, I would argue, is the most heavily regulated industry in the world. Regulations don't solve things." - Wilbur Ross

Governments Use Monetary Policy to Play Santa Claus

It is the nature of democratically elected governments that elected officials must get votes. They do this by giving special interest groups what those groups ask for. The special interest groups, in turn, provide the votes of their members to the politicians that gave them "gifts" from the public coffers.

Over time, people begin to think of the money they receive from the government as a right. What people once saw as a great boon from a generous elected official is now taken for granted and demanded. When the next election rolls around, voters will not even consider a politician who does not support the handouts they are currently receiving. In fact, in the 2012 US presidential elections, large numbers of voters publically threatened to kill candidate Mitt Romney if he was elected because they thought he would cut off their food stamps[42] even though he never said he would.

> "The story of Detroit's bankruptcy was simple enough: Allow capitalism to grow the city, campaign against income inequality, tax the job creators until they flee, increase government spending in order to boost employment, promise generous pension plans to keep people voting for failure. Rinse, wash and repeat." - Ben Shapiro

The amount of money spent on government programs becomes larger because politicians have to create new "gifts" to give voters to encourage new votes. The previous gifts are long forgotten. Therefore, the number and variety of government handout programs increases. Nobel prize-winning economist F. A. Hayek observed:

> If governments are to remain in office in the prevailing political order, they have no choice but to use their powers for the benefit of particular groups–and one strong interest is always to get additional money for extra expenditure. However harmful inflation is in general seen to be, there are always substantial groups of people, including some for whose support collectivist-inclined governments

---

[42] "The 14 Creepiest Threats to Kill Mitt Romney…Over Food Stamps?" http://www.theblaze.com/stories/the-14-creepiest-threats-to-kill-mitt-romney/

primarily look, which in the short run greatly gain by it. (Choice in Currency, F. A. Hayek, p. 42, Ludwig von Mises Institute).

What Hayek is saying is that politicians will always buy votes from powerful special interest groups that can provide them. They will do this by spending public money on those special interest groups. Like junkies turning to crime to sustain their drug habits, politicians will sell out their country's long-term interests in order to serve their own short-term interests. They will do this regardless of the long-term damage they do to the economic lives of their citizens. As long as they have the opportunity to give gifts to special interest groups, they will.

> "Politics is the gentle art of getting votes from the poor and campaign funds from the rich, by promising to protect each from the other." ~ Oscar Ameringer

The money for all this gift giving must come from somewhere. The amount that the US government receives in taxes was long ago exceeded by the amount it spends on its programs. To cover the difference, it borrows money—often from foreign nations. The national debt grows out of control. To keep up with the payments, the government just prints money. Figure 2.5 illustrates how this process works.

When the government prints more money than the market can handle, it devalues the money in your savings account. That value goes somewhere. And of course, that somewhere is right into the money printed by the government. Without decreasing the amount of money in your bank account, the government has stolen from you just as if it had robbed you at gunpoint.

Inflation is a tax that is cleverly hidden from the average consumer. The government uses an army of highly-trained rhetoricians who write endless books, papers, and articles about how necessary inflation is and how we must have it for a healthy economy. This is nothing more than propaganda to get us to accept inflation as a normal course of affairs.

> "Every time the Fed implements 'quantitative easing,' a.k.a. printing more money, two things go up: taxes and inflation. When taxes and inflation go up, more jobs are lost." - Robert Kiyosaki

Once the government prints excess money, it can then spend the money on making payments on the loans it contracted in order to give money to special interest groups. The special interest groups get the big payoff. Or so they think.

1. Politician Wants Votes.

2. Politician Passes a Bill
to Spend on a
Special Interest Group.

3. Government Borrows
Money from Someplace
like China.

4. To Make Payments on
Its Loans, Government
Prints Money.

5. Special Interest Group
Gets the Goodies.

6. Prices Go Up Because
Too Much Money is in
Circulation. Your Money
is Worth Less.

Figure 2.5 Politicians print money to help themselves get votes and give handouts. But it means your paycheck and savings buy less over time.

Surprisingly, people getting government handouts actually believe they are getting something for free. They do not seem to realize that in order for government to give them something, it has to take something away first. And it does. Politicians in the government take away the value of the money in our bank accounts, and then they pretend that they are handing us something for free. The reality is that we *all* pay for *everything* the government hands out. Inflation is universal. It is not selective. It d oes not discriminate. It does not care what color you are. It does not give anyone a break. It has no mercy or compassion for those who are struggling. It hits *everyone* in the economy. No one can escape because of the government monopoly over money.

> "Long before folks fretted the demise of 'quantitative easing,' I fretted its existence. It proved the reverse of its image, an antistimulus, and we've done okay not because of it, but despite it." - Kenneth Fisher

A tricky way the US government has used recently to magically "buy" its own debt is called Quantitative Easing. What happens in Quantitative Easing is that the Federal Reserve buys up the assets of banks, usually the debt of the US government in the form of government bonds and other securities.

The money for this spending spree has to come from somewhere. Fortunately for the Federal Reserve, they have all those nice shiny printing presses that are great at printing money[43]. So that's what they use. The money to purchase all of this debt is just printed[44]. The new money floods the economy causing massive inflation. All prices go up. With Quantitative Easing, so much money is printed that prices went up *much* faster than wages. That means that if you live in the United States of America, you are getting poorer. As long as the government has a monopoly over the currency, there is no escape from poverty for you. No matter how much money you have, the government will eventually use it all up for you. It's just a matter of time.

> "This progressive deterioration in the value of money through history is not an accident, and has had behind it two great driving forces – the impecuniosity of Governments and the superior political influence of the debtor class." – John Maynard Keynes

All governments that have pursued policies like Quantitative Easing have caused hyperinflation and collapsed the value of their currencies. There is no reason to suppose that the same thing will not happen to the US dollar. In fact, we are well on the way to our own eventual collapse, as indicated by Figure 2.6.

As you can see in Figure 2.6, the US has acquired so much debt that there literally is not enough physical cash in circulation in the entire world to pay it back. In fact, most of the realistic options for paying back the debt that ths US has acquired is really just way of shifting one form of debt to another. As the cliché goes, we would just be robbing Peter to pay Paul. The only possible outcome here is the complete collapse of the dollar economy.

---

[43] This is a bit of hyperbole. They actually just use their computers.
[44] Some say that QE is a zero-sum move on the Fed's part because their actions will eventually be reversed. That is, the Fed is not actually "printing money." Not true. Wall Street is being artificially propped up by Quantitative Easing. The market took an immediate downturn when the Fed announced that they would "taper" the flow of new money into the system from $85 billion to $75 billion. Reversing QE and taking all of the new money out of the system would collapse the economy completely. Because of the impossibility of reversing QE, it is no different than just printing new money.

US Revenue, Deficit, and Debt

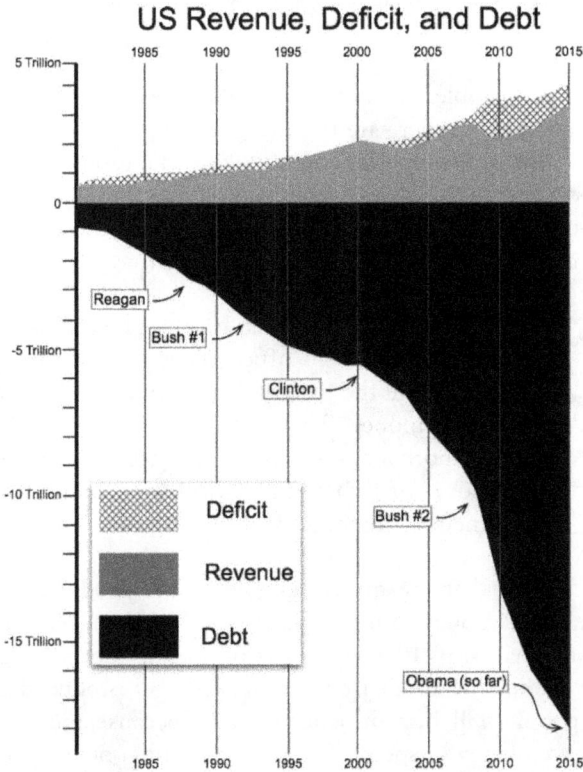

Figure 2.6 America has taken on so much debt that it is impossible to pay it back.

Governments Use Monetary Policy to Temporarily Suppress Unemployment

It is actually possible that overprinting money can *temporarily* help the economy. According to the well-known Phillips Curve, which is used by most economists, inflation can temporarily decrease unemployment.

While decreasing unemployment sounds good at first, it ignores the *reason* for the unemployment. In a functional free market, unemployment normally occurs when there is a change in the market. Specifically, if the market is producing too much of a product or service, the competition drives the price down. That means that some of the individuals or companies producing that product or service can't make enough money to survive. The result is unemployment.

The solution to this problem is that individuals and companies should change and do something else. If there are too many bakers, then some of the bakers should go get a job doing something like cooking in a restaurant, for example. If there are too many companies making hats, some of them should switch to making coats or whatever there is a demand for. This is how the free market solves the problem without government intervention.

> "By a continuing process of inflation, government can confiscate, secretly and unobserved, an important part of the wealth of their citizens."
> - John Maynard Keynes

Elected officials look bad if unemployment goes up while they are in office. They must fix the problem—at least until the next election. An easy solution is to print more money. This makes money less costly to borrow so more loans are made. Businesses invest in producing more of what they already produce. This makes unemployment go back down.

The problem here is that the market is already overproducing in that area. Generating cheap money means that the market will increase its overproduction. That is, if the market was producing too many hats already, printing more money means that even more hats get produced than before. For a while, people will buy the cheaper hats because more of them are working and have money to spend. The price has not gone up—yet.

Overprinting money means that eventually, price inflation will occur. That is, the price of everything will go up. When it does, the supply in the market is even more out of whack with respect to the demand.

To continue the hat example, inflation has ensured that hats are now more overproduced than they were before. And now more people are employed in making hats than before. But inflation (overprinting currency) makes prices go up (price inflation) so people stop buying hats again. Even more people get laid off. The economic impact is worse than if the government had done nothing at all. However, politicians don't see it that way. They panic and intervene once again. The whole cycle repeats itself. This is the origin of the so-called business cycle.

If the government did not cause inflation, then the market would sort itself out. Yes, some hat makers would not be happy for a while. But the market would remain healthy, correct itself automatically, and employment would go up in other areas (such as making coats).

Overprinting money creates a permanent cycle of economic boom and bust that gets worse each time we go through it. The end result can only be complete economic collapse. It doesn't happen immediately, but it does happen. It is unavoidable as long as government has a monopoly on currency.

Government Uses Monetary Policy to "Fix" Trade Imbalances

If a country's goods and services are priced too high, other nations can't afford them and won't buy them. Exports will be lower than imports. The result is what's called a *trade deficit*[45]. Trade deficits make the federal government look bad. Politicians think they can fix this, so they devalue their currency by printing too much of it. This means that everyone else's currency buys more in your country. It has the same effect as forcing businesses to price their goods lower to meet the needs of the government rather than the needs of the free market. Businesses don't actually change their prices. But because everyone else's money buys more in your country, it's just as if the government had set price controls on your products. But price controls never work.

For example, if the value of the US dollar goes down against the euro, then people in countries that use the euro can more easily afford American goods and services. Therefore, they buy more and the politicians look good because they've "fixed" the trade deficit. However, devaluing the dollar eventually makes all prices go up. So exports slow down and the trade deficit returns. And the dollar is now devalued. This can't go on forever. The dollar can only lose so much buying power before it becomes valueless.

> "High inflation occurred in tandem with a sharp recession and this confluence of events was in large part responsible for labor strife and social discord that proved disastrous for the economy and threaten the foundations of the Japanese model of economic cooperation." - Michael M. Hutchinson, Takatoshi Ito, Frank Westermann, "The Great Japanese Stagnation: Lessons for Industrial Countries"

Using inflation to "fix" trade deficits inevitably results in stagnation within the nation. Japan is a perfect example. These days, the Japanese speak of the "lost decade" and the "lost generation" that have resulted from the constant devaluation of their currency and the resulting economic slowdown[46]. The

---

[45] In reality, no such deficit occurs. The economy automatically balances itself out in other ways. Under the current economy, for example, we buy lots of manufactured goods from China. They buy our debt, which is (surprisingly) an actual product in our current system. So it balances out.
[46] There are actually more factors involved in the Japanese situation, all of them relating to poor monetary and tax policy. But inflation has been the most detrimental.

"lost decade" is the 1990's. It's now stretched into two decades of limited economic activity, limited growth, and limited opportunity for the young. It's very common in Japan for people under 35 to be single, living with their parents, and working two part-time jobs. They have become the "lost generation" because they are not moving forward in life.

> "Rising prices or wages do not cause inflation; they only report it. They represent an essential form of economic speech, since money is just another form of information." - Walter Bigelow Wriston

Young Japanese are also not reproducing, so there is another sense in which Japan has lost a generation—the generation that *should* have been born. Japan is on the verge of a population implosion that is directly traceable to the nation's economic policies of long-term inflation. We are beginning to see the same thing occur in the West.

THE CONSEQUENCES OF ECONOMIC MONOPOLY

When government has monopoly power over a nation's currency, there are inevitable consequences. First, we enter a cycle of increasingly larger booms and busts. Second, the government centralizes its control over all aspects of our lives. Third, the economy becomes overly brittle and unable to adapt over time.

Boom and Bust

Whenever government causes inflation, the result is a cycle of boom and bust. Yes, we're talking about the hateful and ridiculous business cycle again.

> "Inflation is as violent as a mugger, as frightening as an armed robber and as deadly as a hit man." - Ronald Reagan

Each time we go through the cycle, the economy becomes worse off and so do we. Debt mounts up and threatens economic stability. Unemployment gets worse with each iteration. The market is unable to correct itself because of the constant pumping by the Fed.

This boom-and-bust cycle is completely unnecessary. It is a direct result of government tampering with the market in the form of inflation. In a free market of competing currencies, the government would not be able to devalue its currency (cause inflation) because if it did, then everyone would dump the dollar in favor of something else. The devaluation would not accomplish its purpose and would result in the collapse of the dollar. To prevent that, the government would have to pursue a saner monetary policy—and a saner fiscal policy as well.

As long as we allow our government to have monopoly control over our currency, we have no choice but to put up with this destructive boom-and-bust business cycle. As soon as we realize that we have alternatives, we can get off this monetary merry-go-round and have an economy that is more stable, easier to prosper in, and more humane than what we have now.

Increasing Government Control

When the boom-and-bust cycle iterates, the typical action for politicians is to blame the rich to obscure the fact that they themselves are the cause of the problem. They then pass laws that give the government more control over the economy so that they can be seen to be doing something and so that they can get more power for themselves. This leads to an increased dependency on the government, which in turn creates a centralization and solidification of their power base.

Calling for more regulation is supremely ironic given the reality of history. After the savings and loan bailouts of the 1980s, there were calls for increased regulation over the entire banking industry. And in fact, every decade since then has seen more banking regulations. But in early 2008, *none* of these regulators saw the mess of subprime mortgages and the impending banking meltdown. All those regulations and all of the regulators were useless in foreseeing and preventing a major economic collapse. How can anyone think that more regulations and more regulators will make the situation any better?

> "Permit me to issue and control the money of a nation, and I care not who makes its laws!" - Anonymous

As centralized control of the economy increases, democracy perishes. No democracy can survive a centrally controlled economy because a centrally controlled economy leads to control over every other aspect of life. In such an environment, political freedoms cannot last.

To emphasize this point, there is no lasting political freedom where economic freedom does not exist. Economic freedom cannot endure if economic democracy does not exist. And an economic democracy requires that government cannot have a monopoly over the creation and governance of currency.

> "The way to crush the bourgeoisie is to grind them between the millstones of taxation and inflation." - Vladimir Lenin

> **Note**
>
> For more information on how inflation and the government's monopoly over currency centralizes its control over all aspects of human life, please see Freedom Under Siege: The US Constitution After 200 Years by Ron Paul, chapter 6 in its entirety. Also, Denationalization of Money: The Argument Refined by F. A. Hayek, chapters 3, 4, and 21. Another good source is The Free Market and Its Enemies: Pseudo-Science, Socialism, and Inflation by Ludwig Von Mises, chapter 6.

THE REAL EFFECTS OF INFLATION

Some economists say that printing more money is a good thing. In fact, it's their answer to every economic problem. Figure 2.7 illustrates their thought process.

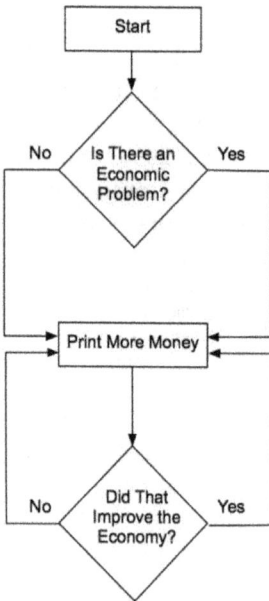

Figure 2.7 Some economists think that the answer to *everything* is to print more money.

As nutty as the thought process shown in Figure 2.7 is, there are leading economists who actually think that steadily inflating the dollar into worthlessness is a good thing. In fact, the Federal Reserve Bank states on its web site that its goal is a steady inflation rate of 2%.[47]

Let's do some math.

If you start with $1 and have an inflation rate of 2% per year, your $1 will hit a value of zero in 229 years. To be more precise, after 229 years it hits a value of less than one penny.

But let's think, will people really keep confidence in the dollar as its value draws closer to zero? Or will they panic when the value of the original $1 falls below 10 cents after 115 years?

---

[47] "Why does the Federal Reserve aim for 2 percent inflation over time?", The Board of Governors of the Federal Reserve System, http://www.federalreserve.gov/faqs/economy_14400.htm.

Now let's remember, the dollar has been under the management of the Fed for just over 100 years. In that time it's lost 98% of its original

> "Having a little inflation is like being a little pregnant." – Leon Henderson

value. And we're going to *continue* to inflate the dollar at 2%? The only possible result of continuing to decrease the value of the dollar is that one day the dollar will be worth nothing. There is no other way this can end.

The Government is Stealing from You

Inflation is actually a tax. It's a way for the government to take money from you without taking money from you.

Say what?

Here's how it works. Although I've touched on this briefly already, let's use an example to clarify how inflation really works.

Suppose we all live in a small village together on a tropical island. Taken as a whole, the village produces all that we need. We all decide to use a kind of rare clamshell as our money. The village Elders hand out 10 clamshells per family. There are 10 families in the village, so there's a total of 100 clamshells.

One day I discover a new source of clamshells. No one else knows about this. "Ha ha!" I exclaim. "I'm rich!"

I gather 10 clamshells and rush back to the village. I promptly spend them on a new pair of sandals.

There are now 110 clamshells in circulation. But remember that the amount of goods and services available in the village hasn't changed. Only the amount of currency in circulation has gone up. Goods and services are the real wealth, not money. Simply creating more money does not create more wealth.

Putting more clamshells into circulation means that the value of each individual clamshell has gone down. I have inflated the currency. In doing so, I've taken a little bit of the value out of each clamshell that you possess.

Where does that value go? It goes right into the extra 10 clamshells that I just spent. By inflating our clamshell currency, I've taken value out of your money and put it into mine. Then I spent that value on something nice for me.

In other words, I just stole from you. And I didn't have to mug you or sneak into your hut to take your clamshells. I used money as a proxy for theft.

Not only that, I just got myself a nice, new pair of sandals that would have gone to someone else in the village. So the production of sandals has to go up if everyone is to have some footwear. Either that, or someone is going to have to do without.

This little story shows exactly what inflation does in our economy. It steals value from your bank account without decreasing the number of dollars that you hold. And where does that value go? It goes into that nice, new money that the government just printed.

Like the village in our example, the amount of goods and services in the economy doesn't go up just because the government prints more money. Either the economy has to expand so that everyone has enough or some people have to fall into poverty. The reason that they have to fall into poverty is that the government used inflation to rob us all in a very quiet, simple way. All they while they steal from us, they tell us how wonderful and necessary inflation is.

Does this make sense to anyone at all?

Retirement? What Retirement?

Since the 2008-2009 meltdown, the government has been reporting inflation at 2%-4%.

This is a lie.

This rate is based on the Consumer Price Index (CPI). The government calculates the CPI by monitoring the prices of a group of selected (by the government) goods and services. They call this group of products and services the CPI's *basket of commodities*. Unfortunately, the government manipulates the contents of this basket so that it results in a CPI that makes the government look good[48].

---

[48] Perianne Boring, "If You Want To Know The Real Rate Of Inflation, Don't Bother With The CPI", Forbes.com, 02/03/2014, http://www.forbes.com/sites/perianneboring/2014/02/03/if-you-want-to-know-the-real-rate-of-inflation-dont-bother-with-the-cpi/ and John Melloy, "Inflation Actually Near 10% Using Older Measure", Cnbc.com, 04/12/2011, http://www.cnbc.com/id/42551209. See also Fred Caifosh, "Why The Consumer Price Index Is Controversial", Investopedia.com, http://www.investopedia.com/articles/07/consumerpriceindex.asp.

For example, did you know that the CPI *does not* include food or gas? The government says they don't include food and gas in the CPI because "the prices are too volatile." So even though the prices of food and gas more than doubled after the 2008-2009 meltdown, it was not reflected in the CPI and the government glowingly reported an inflation rate of around 3% on a regular basis.

But let's just imagine that the government isn't lying to us. Let's say that the inflation rate really is 3%. Now suppose you have $200,000 in savings. Imagine that you are going to retire in 20 years. You want to use your $200,000 to buy a nice house for your golden years.

Remember that inflation decreases the buying power of your money. At a 3% inflation rate, how much buying power will your $200,000 lose over 20 years?

Well, let's do the math. A loss of 3% of the buying power of you money per year means that at on December 31st, your money has 97% of the buying power it had on January 1st. So we simply multiply $200,000 x 0.97. That shows the decrease in buying power over 1 year. To see how much it decreases over 20 years, we repeat that calculation 19 more times. If you perform this calculation, your $200,000 becomes $108,758.87. In other words, your money now as 54% of its original value. It has lost 46% of its buying power[49].

Look around at the housing market. What can you buy for $108,758.87? I can guarantee that whatever you end up with is not going to be nearly as nice as what you can buy for $200,000. Perhaps you should just lower your expectations for your retirement. Doesn't that sound wonderful? I didn't think so.

In an environment where your buying power is constantly being eroded by inflation, it's next to impossible to provide yourself with a decent retirement. More and more people who are retirement age are continuing to work because they're literally afraid that the government will rob them of their savings. These fears are not unjustified.

Rewarding Businesses for Failing

Here's a quick question: When does the government release new money into the economy?

---

[49] These numbers are rounded. The actual numbers are $108,758.86858534945148, 54.379434292675% and 45.620565707325%.

Answer: The government usually releases new money when the economy is doing badly.

We already saw that the government inflates our currency for three reasons: 1) to hand out "freebies" to special interest groups and thereby gain votes, 2) to "fix" unemployment, and 3) to "fix" perceived trade imbalances.

Number two of these three reasons that the government releases new money occurs when the economy is doing badly. In other words, the government is rewarding businesses with cheap money for doing the wrong thing. Businesses then use that cheap money to do more of what they're doing wrong.

This is an important point that we'll revisit later. In a functional economy, the supply of currency would always expand or contract to match the demand for currency. So new currency should be released when the economy is expanding, not when it is contracting or doing badly as our current system does.

In a functional economy, the price of goods and services sends a signal to producers to produce more or less of those goods and services. By inflating our currency when the economy is doing badly, the government distorts prices and gets our economy to spend money in all the wrong places. In other words, the economy would heal itself faster and better if there were no inflation.

> "If you wish to destroy a nation, you must first corrupt its currency" – Adam Fergusson

By rewarding businesses for doing the wrong thing, the government is hobbling our economy and keeping it from reaching its maximum potential. There are actually fewer jobs in the economy now than there could be if there was no inflation.

THE ECONOMY, NETWORKS, AND "TOO BIG TO FAIL"

The recent emergence of banks that are considered "too big to fail" demonstrates that our economic system has become overly dependent on a few of the major players. This is a recipe for failure on a scale that can literally end human civilization.

The Economy is a Network

Our economy is a network of entities that interact via currency-based transactions. As such, it operates in exactly the same way that any other network operates. This includes computer networks. The same principles that apply to networks of computers apply to our economy. This is fortunate because the operation of computer networks is well understood.

Resilient Networks

When designing a computer network, systems analysts must strike a balance between efficiency and resilience. They create that balance using the number of connections between computers. Figure 2.7 gives an example.

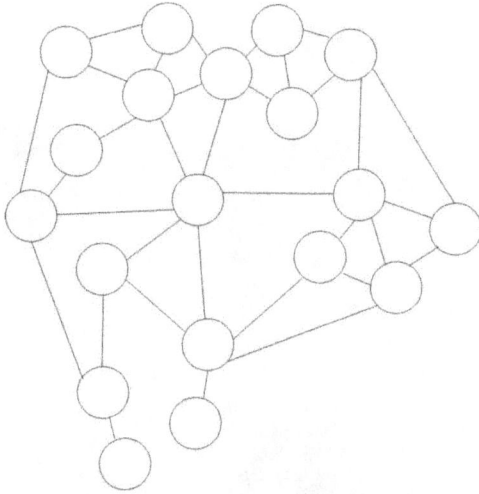

Figure 2.7 A Network that is Resilient but Not Efficient

In Figure 2.7, the circles represent computers in the network. The lines between them represent network connections between the computers. As you can see, this network is highly connected. That makes it very resilient. Even if large parts of the network become inoperative, the remaining computers can still communicate with each other. So the strength of this network is that it is nearly indestructible—especially if the computers are spread out across the world.

The downside of this network is that it is not very efficient. When sending data between computers, the network can spend an inordinate amount of time finding the shortest path from one computer to another. The network is resilient but not efficient.

Brittle Networks

Now let's take a look at a different network.

The network shown in Figure 2.8 solves the efficiency problem of the network in Figure 2.7. As you can see, the network in Figure 2.8 has far fewer connections. Communication from computer to computer takes place in a very efficient way.

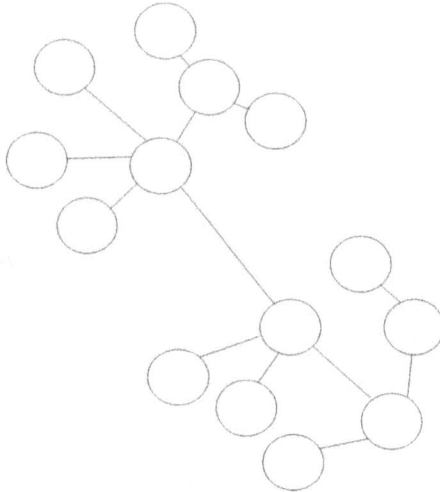

Figure 2.8 A network that is efficient but not resilient.

However, the problem here is that the network is too brittle. One bomb (or one earthquake or one tidal wave) anywhere along the main trunk line takes down the whole network.

Striking a Balance

The balance between efficiency and resiliency was a key factor in the design of the Internet. The Internet is specifically designed to keep functioning even if part of it is literally hit with a nuclear bomb. It is also designed to be efficient enough to maintain the pace of modern business.

All networks of all types function according to this same pattern. It doesn't matter whether the network is a worldwide economy or the ecosystem of the Florida everglades. They all work this way. They all have what's called a *window of viability* where they have a stable balance between resiliency and efficiency. If they operate outside of the window of viability, they are likely to self destruct.

The Economic Network

In our economy, the circles in the two previous figures represent businesses. Businesses perform transactions. Without transactions, they fail. Money provides the connections that make the transactions possible. In the diagrams, the lines connecting the circles represent money.

In our economy, all transactions have to pass through the banks and credit card companies. If something happens and the banks and credit card companies fail, the transactions can't take place. This makes businesses fail, even if the businesses have customers who want to buy their goods and services.

Our economic network can't function without our transaction processors (banks and credit card companies). Using the banks and credit card companies is very efficient. But we're so dependent on them that we can't let them fail. In other words, our economy is too brittle and it may collapse as a result.

To help make this a little clearer, let's look at a quick example. Suppose you own a business that produces machinery for factories. Each factory has to have its own type of machinery. That is, chewing gum factories need you to build one type of machinery for them while hair gel factories need other types of machinery.

> "As a young man, I lived through the Great Depression, when banks failed and so many lost their jobs and homes and went hungry. I was fortunate to have a job at a canning factory that paid 25 cents an hour." - James E. Faust

Now imagine that you get an order from a customer. That means that you have someone ready and willing to pay you as soon as you design their machinery, build it, and deliver it.

Under normal circumstances, you go to your bank and get a short-term bridge loan. You use the loan to pay your suppliers and your employees. When you complete the order, you get paid and you use the funds to pay off the loan. All is good so far.

Imagine that the overall economy is doing badly but there still is demand for your company's services. Because the economy is not doing well, banks have tightened their lending policies. You cannot get a bridge loan. Other than the banks, there is no one that you can go to for your bridge loan. If there were, you could complete your transaction and stay in business. Since there is not, your business may well fail even though it has an order that will keep it afloat through the economic downturn.

Depending too heavily on one group of transaction processors, in this case the banks, means that transactions that could and should take place will not happen if something goes wrong in the economy. The business in this example could not obtain the short-term liquidity that it needed to keep going. If it could turn to an alternate source of short-term capital, then it could fulfill the order and continue to operate.

This is a small example, but our whole economy suffers from this problem. The fact that we have banks that are "too big to fail" means that our economy is overly dependent on too few entities that process or enable transactions. It's like making an Internet that has too few communications servers on it. If that were to happen, too much traffic would have to be routed through two few connections. If those few connections fail, the whole network collapses.

We also depend on too few currencies. Money is what provides the connections between businesses. If a national currency fails, every business within that country will fail. That's bad. But if, for instance, the US dollar fails, it's worse than bad. The US dollar is used to back some 40-60 currencies worldwide. In other words, the currencies of those countries have value because the US dollar has value. If the US dollar fails, then *all* of the other currencies dependent on it will fail. That will set off a domino effect that will destroy the economies of every country on earth.

Even if the national currency doesn't fail, our money has to be routed through banks, credit card companies, and so forth. They process and enable transactions. They provide liquidity where needed. But the bank-based economy has become so efficient that there are very few transaction processors left any more. This is a problem. When just one transaction processor fails, it breaks the connections between large parts of the economy. A domino effect occurs. Transactions that would take place if they could be processed don't occur. Insufficient capital, insufficient access to bridge loans, and insufficient long-term credit all mean that businesses simply can't do business. We saw this demonstrated very clearly in the financial meltdown of 2008-2009.

Some Must Win, Some Must Lose

In a functional economy, we must have a diverse set of transaction processors so that the failure of some businesses won't bring down the whole economy. Banks that are mismanaged must fail. But if our economy had alternate transaction processors, that wouldn't be a problem.

Some say that the answer to this is to bail out the banks. That just rewards them for bad behavior and drains money away from the rest of the economy. It makes us unequal before the law if the banks get bailouts that the rest of us don't. It puts them in a privileged position in our economy.

Won't letting businesses fail hurt the economy?

Yes and no.

Businesses that are mismanaged, that misspend, or that do not meet customers' needs *should* fail. They should not be bailed out no matter what. If they are bailed out, they will be a drag on the economy. Specifically, the economy will not be able to achieve greater levels of prosperity until those businesses are allowed to fail.

Our economy is currently in a state where we have cut ourselves off from economic success because we will not accept the failure of some of our biggest banks. We are in a state of economic stagnation precisely because the "too big to fail" banks were bailed out. Those banks are still doing things that are wrong. And they are dragging down our economy, slowing growth, wasting money, and cutting us off from the possibility of prosperity. Until those banks fail, we will not see much improvement in our economy.

> "Every government interference in the economy consists of giving an unearned benefit, extorted by force, to some men at the expense of others." - Ayn Rand

The Great Resession as an Example

Let's recall why we got into 2008-2009 mess in the first place. Many years ago, Congress passed the insanely stupid Community Reinvestment Act (CRA) that was an attempt to turn banks into social welfare agencies. It forced banks to provide home loans to poor people who could not possibly afford to pay the loans back.

At first, the number of loans given under this program were few, so the law had little negative effect on the economy. As the years went by, however, the government forced increases in the number of these loans–this was especially true under the Clinton administration. This created an artificial demand in the housing market that started a speculative bubble.

As with all speculative bubbles, everyone wanted to get on the bandwagon. Banks began lending willy-nilly and they were roundly encouraged by the government. But predictably, the gravy train eventually derailed. Or to repeat another cliché, the piper had to be paid.

People had borrowed money to buy houses they could not afford. Near the end, many homebuyers were purchasing houses on interest-only loans. That means that their monthly payments were only payments in the interest on the loans, not on the loans themselves. In other words, they would never pay the house off.

Because homebuyers could not pay their loans, they began to default. And the number of defaults grew. The government stepped in and bailed out some lenders and not others. At first, banks saw the government bailing out lenders so they didn't take the actions they should have to protect themselves from failure. Then they saw the government *not* bail out some major lenders. Because those lenders didn't take action in time to keep themselves solvent, they went belly up. And these were "too big to fail" lenders. Our economy depended too heavily on these lenders. Without them, it could not function properly.

The failure of some of the biggest financial institutions in our economy caused a domino effect. Because it became next to impossible to get a home loan, housing prices plummeted even though there were still people who could afford to buy houses.

"Widespread and broad-brush government policies that encouraged greater investment and consumption of housing (including support for Fannie Mae and Freddie Mac) contributed to the housing boom, as did excessively expansionary monetary policies over 2003–2004, when the Federal Reserve continued for too long to worry about the potential of deflation and the adverse effects that would follow."
— Lawrence J. White

Banks tightened *all* of their lending requirements. People couldn't get loans to buy cars, so car dealerships and car manufacturers couldn't sell cars. Quite stupidly, the government stepped in to bail out the car companies by taking ownership of them at a cost of millions upon millions of taxpayer dollars. That *still* didn't stop Detroit from going bankrupt.

People lost their jobs in industries that depended on liquidity (short-term loans) from banks. Because those people weren't working, they didn't spend money on other products and services. That in turn caused slowdowns in other industries. And so the meltdown spread to all sectors of the economy.

This entire–and entirely preventable–fiasco came at a terrible cost. For example, there were actually owners of car dealerships who committed suicide when the businesses their families had owned for generations went bankrupt. This is a hideous cost for what began as a misguidedly stupid attempt to help the poor. And it serves as a gruesome lesson of the effects of government meddling in the economy.

If the government had not passed Community Reinvestment Act in the first place, none of this would have happened. If the government had not intervened and saved some banks and not others, the banks would have done what it takes for failing businesses to survive. That is, many of them would have passed through bankruptcy and reorganized. Or they would have failed outright and their assets would have been bought by banks that were managed better. There still would have been a slowdown, but it would not have been so disastrous. And the failure of some major banks would have created a business opportunity for new types of transaction processors and capital providers (people who make loans) to appear. The result would have been a healthier economy that would be more prosperous by now.

By letting our economy turn into an overly brittle network of "too big to fail" transaction processors that we will bail out no matter what, we have cut some businesses off from failure. And we have cut our entire economy off from success. Businesses that are not allowed to fail when they should are a drag on the economy and our economy will not succeed until they are gone.

> "Government's view of the economy could be summed up in a few short phrases: If it moves, tax it. If it keeps moving, regulate it. And if it stops moving, subsidize it." - Ronald Reagan

Resiliency in the Economy

To increase the resiliency of the economy, we need to diversify it. But in diversifying the economy, it is entirely possible that it will become less efficient. Normally, this would be a cause for concern. However, in this case, less efficiency is desirable. Lowering the efficiency of the economy will improve its resiliency. Transactions will occur that would otherwise not occur.

Recall that all networks have a window of viability. A viable economic network is resilient enough to survive economic change but efficient enough to keep costs down. Our economy is extremely efficient, but the busts in the ongoing boom-and-bust cycles are increasingly shattering our economy's ability to function and slowing healthy growth. [50]

In other words, increasing the number of possible connections in our economic network will actually cause healthy economic growth. We increase the number of connections by creating new forms of money and new ways to process that money. In doing so, we enable transactions to take place that otherwise would not happen. The result is an increased transaction flow and increased economic growth.

In addition, the ability to not depend solely on banks for needed liquidity has a stabilizing effect on the economy. The ups and downs are smaller because more people stay employed and more businesses continue to function. If people had other currencies to use and other ways to process transactions, they could choose whatever currency or transaction method was best for the given economic situation. This would enable more businesses to continue to function even in economically difficult times.

NEITHER HUMANE NOR VIABLE

The current system of banking and money creation that we use is inhumane and self-destructive. Its effects are summarized as follows.

- It forces increasing numbers of people into poverty no matter how well they manage their money.

- It concentrates wealth from the many to the few.

- It creates destructively unsustainable economic growth.

- It drives overconsumption, which in turn devastates the environment.

- It stifles economic diversity.

- It devalues our money and makes it impossible to save.

- It causes cycles of economic booms followed by economic busts. The cycles get bigger each time we go through them.

---

[50] For more information on the economy as a network and network efficiency verses network resiliency, please see New Money for a New World, by Bernard Lietaer and Stephen Belgin, chapter 12 in its entirety.

- It consolidates control into the hands of the government.

- It creates an economy that is so brittle that it is no longer viable. Any economic disruption causes massive repercussions across the entire nation.

Now before we end this chapter, I'm going to remind you of a new technology that I keep mentioning but haven't explained yet. Yes, I'm talking about the blockchain again.

If you bought this book to learn more about the blockchain and what it can do, you will soon be rewarded. In another few chapters, I'll provide a solid look at the blockchain and how it can solve the problems presented so far. I promise you it will be worth the wait. The blockchain enables anyone to create currencies, transaction processors, and financial instruments that cannot currently be created. It enables us to build an inexpensive economic network that can be customized to any degree desired for maximum efficiency.

I keep making BIG claims about what the blockchain can do without providing much detail. That's because right now I'm just presenting an overview of the problems we're facing and pointing out where the blockchain can help. I'll back up *all* of these claims and give even more examples of the power of the blockchain and its derivatives, such as smart contracts and side chains. For now, don't worry that my blockchain claims seem like tall tales or that you can't see how such things could possibly be true. After I've laid the groundwork, all of the information about the blockchain that I give you will make sense.

And if you bought this book trying to find out what the root causes are of most of the problems that face our civilization, you too will be rewarded. In the next few chapters, I'll explain how our government and money systems are *supposed* to work and why the answers we keep trying to use will never solve our problems. Then we'll do a deep (but non-technical) dive into this new blockchain technology that everyone's talking about. So let's keep going in Chapter 3 by getting an understanding of what the government should and should not be.

3 THE GOVERNMENT IS A PAIR OF PLIERS

One of the reasons our government, and consequently our economy, is so dysfunctional is that most people today don't understand what government is and what it's supposed to do.

THE RIGHT TOOL FOR THE RIGHT JOB

Have you ever done any work on your house or yard? Have you ever talked with a professional handyman?

A real handyman will say, "The right tool for the right job." What does this mean?

Professional handymen often see botched repairs done by homeowners who thought they knew enough to do it themselves. A startling number of these bungled repairs involve a pair of pliers.

For some reason, people who are not experienced in home repairs seem to think that whatever you're doing, a pair of pliers is exactly the right tool for the task at hand. The inexperienced will use pliers to tighten or loosen nuts and bolts instead of wrenches. They'll try to turn screws or pull nails with pliers. I've even seen people hammering nails with pliers.

"Unfortunately the Federal Government has strayed far afield from its legitimate business. It has trespassed upon fields where there should be no trespass. If we could confine our Federal expenditures to the legitimate obligations and functions of the Federal Government, a material reduction would be apparent. But far more important than this would be its effect upon the fabric of our constitutional form of government, which tends to be gradually weakened and undermined by this encroachment." - Calvin Coolidge

The point here is that pliers are not the right tool for any of those jobs. But there *are* things that pliers are the proper tool for. So pliers should be used when appropriate.

The government is a pair of pliers. There are a few things that the government is good for. And the government should be used for those tasks. But things that the government is actually good at are very few in number. In most cases, there are far better solutions. Some things that the government should *not* be used for include (but is not limited to) the following.

- Managing the economy.

- Helping the poor.

- Augmenting, defining, or replacing the family.

- Auditing, censoring, or controlling citizen's speech and other communications.

- Defining what are or are not acceptable religious beliefs.

Every effort the government makes in these areas (as well as many others) only backfire and decrease the quality of life for those it supposedly intends to help.

TOO MANY LABELS

In politics today, we tend to think of silly and unrealistic stereotypes and labels. We talk of "liberals" (who are not liberal at all by the real definition of the word) or "conservatives" (what do they conserve, exactly?), as well as "Left" and "Right", Republican and Democrat and Libertarians, and even "libertarians."[51]

None of these labels have anything to do with reality. Why? Because they have no real meaning. Or rather, they mean whatever is expedient for them to mean right now.

For instance, there was a time in the US when the Democratic party, the party of today's supposed "liberals" was the party of racism and the Ku Klux Klan. For example the Democratic Governor of Alabama, George Wallace, stood in the doorway of the University of Alabama to prevent the all-white university from being integrated with two black students. In a hilarious rewrite of history and a mind-bending relabeling for convenience, US News and World Report dubbed Wallace, a supposedly liberal Democrat, a "staunch conservative."[52] Wallace himself most likely would have vomited at being

---

[51] Apparently some people think there's a difference between a Libertarian and a libertarian. I have no idea what the difference is.

[52] Debra Bell, "George Wallace Stood in a Doorway at the University of Alabama 50 Years Ago Today", US News and World Report, June 11, 2013, http://www.usnews.com/news/blogs/press-past/2013/06/11/george-wallace-stood-in-a-doorway-at-the-university-of-alabama-50-years-ago-today

called a conservative. And if not, does that mean that the Democratic party was once the party of conservatives? I think modern Democrats would hate such a statement.

Likewise the Republican party used to bill itself as being the party of the common man. There was a time when most Republicans were members of a labor union. And lest we forget, Abraham Lincoln–the man who issued the Emancipation Proclamation that freed all slaves in the US–was a Republican president.

Today, by all accounts, everything is exactly the opposite. It's all for convenience. The labels of the past are recast to mean whatever is expedient for them to mean today. Political parties' philosophies change over time until the party represents exactly the opposite of what it used to.

THE FOUNDERS' VIEW

When the US Constitution was written, the Founders had no conception of the pointless labels we used today. They thought in terms of balancing the liberty of the individual against the needs of the nation. For them, striking this balance was critical.

ANARCHY

If the government ended up being too focused on the liberty of the individual, then chaos and anarchy would result. There are those who believe we can live in an anarchic state and still have order. This is a naïve world view. There are some things that the government just has to do and the Founders understood this well.

Finding the appropriate amount of individual rights is a tricky thing to do. Exalting the needs of individual over the needs of the society that individuals live in can end up with a non-functional civilization.

TYRANNY

On the other hand, if the government was too focused on the needs of the state, then tyranny would result. The government would eventually criminalize virtually all human behavior and have arbitrary powers to lock up or execute people at will.

We're already arriving at the point where nearly every behavior is a criminal offense. In December of 2014, the New York City government discovered that Eric Garner, an overweight, asthmatic, small-time lawbreaker, was selling cigarettes on the of New York City. This doesn't seem to be criminal activity, but one has to know that the mayor of New York, Michael Bloomberg, has decided that he has the right to dictate what size soda you can drink and whether or not you should smoke. So he imposed a cigarette tax to discourage New Yorkers from smoking. New Jersey, just a short bus ride away, doesn't have this tax. Eric Garner, an African-American, was buying cheaper cigarettes in New Jersey and selling them on the streets of New York. He did not have a business license, which means he did not ask the government's permission to engage in economic activity. He was not paying taxes on those sales and heaven help anyone who stands between Mayor Bloomberg and the taxes he thinks he deserves.

When Eric Garner was found selling cigarettes, a group of police officers tried to arrest him. In the process, one of the officers put Mr. Garner into a chokehold that was banned by the New York City police department. Mr. Garner's last words as he was pinned to the ground were, "I can't breathe." A medical examiner ruled that Mr. Garner died of "compression of the neck."

Whenever the government has criminalized any behavior, there is always the possibility that people will die as a result. Legislators, city councilmen, and mayors don't seem to get the fact that their attempts to control others' behavior usually result in the deaths of people like Eric Garner. Was collecting a few taxes really worth a man's life? And were the taxes imposed by a mayor and city council seeking to stop people from smoking really the right thing for a city government to do? Clearly not.

Government *always* has a reason why it needs more power. This is particularly true of law enforcement. They want to make their own jobs easier. So they expand their surveillance powers over the general population for what they think are very good reasons. We've seen this repeatedly since the passage of the hateful Patriot Act[53] (which has nothing to do with patriotism and everything to do with control) by the George W. Bush administration.

---

[53] In spite of enabling the government to spy on every American, the FBI has admitted that the Patriot Act has helped it crack exactly zero terrorism cases. Maggie Ybarra, "FBI admits no major cases cracked with Patriot Act snooping powers", The Washington Times, May 21, 2015, http://www.washingtontimes.com/news/2015/may/21/fbi-admits-patriot-act-snooping-powers-didnt-crack/.

Likewise, the government has an endless river of reasons why it needs greater control over the economy. But in the end, allowing it leads to tyranny.

Figure 3.1 illustrates the Founder's view of government.

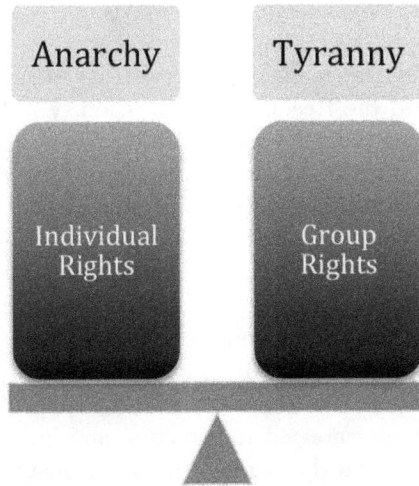

Figure 3.1 The Founders saw the need to balance
the rights of the individual and the rights of the group.

In the eyes of the founders, too many individual rights would lead to a nation that wasn't cohesive enough to function; it would create anarchy. Likewise, granting the group, in the form of the government, too many rights would lead to tyranny.

What the US Constitution was built on was the idea of balance. That's why we use the term "checks and balances" when we discuss the Constitution. The main focus of the Constitution is to provide balance. So it granted specific powers to the federal government and left most other powers with the states. It limited the power of the states and granted all other powers to the individual.

In all of this, there is no concept of conservative, liberal, libertarian (and Libertarian), Left, Right, Democrat, or Republican. These are simply labels that we do not need. They only serve to divide people rather than help them find a balance between anarchy and tyranny.

LIMITED GOVERNMENT

The Founders saw limited government as the most effective way to create the balance between anarchy and tyranny. This was the primary reason for adding the Bill of Rights before the Constitution was passed.

For example, they understood that in order for the balance to be maintained, the people would need the freedom of expression. When Thomas Jefferson was elected president in 1801, he said in his inaugural address that, "the will of the people is the only legitimate foundation of any government, and to protect its free expression should be our first object."

In other words, the government derives its powers from the people. They grant powers to the government because the people have what the founders called "natural rights" and the government doesn't. So the government can only do what the people allow it to do.

The US Constitution was created to structure and limit the powers of the government. It doesn't limit the rights of the individual at all. It just gave the society at large, as represented by the government, enough of the people's power to maintain the balance between liberty and anarchy.

THE INTERNET AS AN EXAMPLE

The idea of limited government was treasured by the Founders because they knew that the American people can accomplish near miracles if only the government would keep out of the way. And the best example of that is the Internet. So let's just take a look at the Internet and see how it got to be the proverbial best thing since sliced bread.

MYTH: THE GOVERNMENT CREATED THE INTERNET

It's a common myth that the government created the Internet, and therefore the Internet is a perfect example of how fantastic big government and government-funded projects are. This could not be further from the truth.

However, this myth is so common that Barak Obama used it in a speech when he said, "The Internet didn't get invented on its own. Government research created the Internet so that all the companies could make money."

Mr. Obama's statement is so completely wrong that it's outright laughable.

Even if we were to assume that the government really *did* create the Internet (it didn't), the government *did not* create the Internet so that companies could make money. It created a decentralized network for its own uses—primarily to connect defense contractors to the Department of Defense. At no point in history did the government ever worry about companies making money off of this network. In fact, they did their best to prevent it.

> "It's important to understand the history of the Internet because it's too often wrongly cited to justify big government." – L. Gordon Crovitz

Arpanet was Not the Internet

It's true that the government did create Arpanet, one of the earliest distributed networks. But Arpanet was never a big player in the development of the Internet. By 1972, Arpanet linked just 37 computers in the whole world.

And it's also true that Arpanet did introduce TCP/IP, the underlying transport protocol that the entire Internet uses. But even then, the inventors of Arpanet and TCP/IP relied heavily on work done at private companies. For example, neither Arpanet nor TCP/IP could have succeeded without the invention of Ethernet connections by Robert Taylor at Xerox Park labs.

If we're giving credit where credit was due, then Xerox PARC should rightly be credited as the inventors of the Internet. Bell Labs should also be credited for creating Unix, the operating system that has long been the backbone of the Internet. Unix was the first operating system to incorporate TCP/IP, and thus was the first operating system to really be networkable on any commercial scale[54].

Xerox and Bell Labs Miss the Boat

If Xerox and Bell Labs invented the Internet, then why aren't they big Internet companies today?

There are several reasons for this. The first is that they didn't understand what they had. Xerox was focused on selling copiers. They created Ethernet so that people could send documents to printers from their computers. It

---

[54] There are some who would dispute this. Most notably, IBM was working on networking computers at this time.

simply didn't occur to them that they could create a decentralized network that would develop into what we know as the Internet today.

Likewise, Bell Labs didn't see the potential of Unix. It came from a failed operating system called Multics. At one point, Bell Labs actually forbade anyone to do any further work on Multics and any related projects. But an engineer by the name of Ken Thompson said, "I'm going to work on it anyway." He quietly created Unix without anyone knowing[55]. Eventually, Bell Labs saw that Thompson had created something good, but they didn't see the commercial potential for it. They made it available for universities to use and even provided the source code for it for free. That's how it became the backbone of the Internet.

The Government Delayed the Development of the Internet

Another reason that private companies like Xerox and Bell Labs didn't build a large commercial Internet was that the government prevented private companies from establishing distributed commercial networks. The FCC regulated *all* communications under a set of laws called Title II. And the regulatory barriers that the FCC erected made it commercially unviable to create an Internet.

So it was the government's interference with the development of private commercial networks that prevented the real Internet from being born. Like Xerox and Bell Labs, the government had no real idea what it had.

For a long time, the Internet was the quiet purview of defense contractors, government agencies, and universities. This was the direct result of government interference. The government actually prevented the Internet that we see today from being born for about thirty years.

"The Internet, in fact, reaffirms the basic free market critique of large government. Here for 30 years the government had an immensely useful protocol for transferring information, TCP/IP, but it languished. . . . In less than a decade, private concerns have taken that protocol and created one of the most important technological revolutions of the millennia." – Brian Carnell

It's worth noting that there *was* a strong demand for an Internet-style network in the private sector. But the inability of private companies to provide it due to government interference led to awkward workarounds. Does anyone remember Compuserve? Or America Online[56]? These companies

---

[55] This information was related to me in my university days by one of my Computer Science instructors, who worked at Bell Labs with Thompson and repeatedly cautioned Thompson not to get caught working on Multics.
[56] During this time period, neither Compuserve nor America Online were connected to the Internet.

provided dial-up[57] connections to their private in-house network of computer servers[58]. And those servers enabled people to put up stores, have chat rooms, and a lot of the other things we see on the Internet today.

The only reason services like Compuserve and AOL even existed was because the government was preventing the birth of the real Internet.

DEREGULATION MADE THE INTERNET AWESOME

Over the decades, the government decided it didn't want to handle the burden of the Internet. So it deregulated the Internet and privatized it. Importantly, the government reached an agreement that its internet and internets established by commercial companies would freely pass each other's network traffic through their own networks. This let almost any company connect their computers to what we now know as the Internet.

The government also decided that the FCC would no longer have the power to regulate the Internet under Title II. Once the Internet was fully privatized in 1995, it simply exploded. The only reason we have the Internet that we have today is because of deregulation and privatization. The Internet is the ultimate proof of how stifling government regulation is and how empowering deregulation is.

By getting the government out of the way and allowing private companies to build whatever they wanted, we got Twitter, Google, Facebook, and so very much more. But this would not have been possible if government hadn't deregulated the Internet. Government simply can't predict innovation. The people in government have nothing that pushes them to innovate. They have no way to see how technologies they have can be used for awesome things like the Internet that we know and love.

> "Yes, President Obama, government invented the Arpanet. But what happened next shows how government fails, but individuals succeed. Government enacted barriers to private-sector research, and took decades before it allowed all of us to benefit from an important new technology. Once it was privatized, individuals - not government - created the internet that we know today." – John Stossel

Simply put, if it hadn't been for deregulation, we would all still be using American Online. And not the America Online we see today. The old one;

---

[57] Dial-up is done with modems. Beeeeeep boooop kzxrtvzzz, and so forth. The younger generation has no idea what this means.
[58] Private in-house networks were not regulated under Title II at the time. Commercial internets were.

the one with dial-up modems[59] running at 56,000 bits per second. We would not have always-on broadband connections running at 1 megabit (or 10 megbits or 25 megabits) per second in our homes.

So while it's true that the government invented Arpanet and Arpanet contributed to the development of the Internet, it was the government that *prevented* the real Internet from being born. Once the government got out of the way, the Internet became every kind of awesome imaginable.

Now some people will still insist, "The government invented the Internet. No Arpanet, no Internet."

False. It's clear from the histories of online companies that there was a demand for the Internet as far back as the early 1980s and possibly as early as the 1970s. If they could have, private companies would have been glad to step up and provide us with a real Internet. But government regulations prevented any networks besides Arpanet from appearing. So if we want to make any concessions to these people (we don't) we'll say that if "no Arpanet, no Internet" is true, then it's only because the government *forced* it to be that way. If the government had gotten out of the way sooner, then it's obvious that Arpanet would not have ever been needed to get the Internet started.

THE FCC IS DESTROYING THE INTERNET

Now let's look at what's happening today. Recently the FCC has decided it owns the Internet (Free speech? What free speech?). With no congressional authority whatsoever[60], the FCC has simply declared itself the regulator of the Internet. And once again, they want to use Title II to do that.

This time, they're saying they're "protecting" us from the big bad Internet service providers (ISPs). This is like mobsters coming into your business and selling you "protection" from any "unfortunate accidents" that might occur.

Title II will Kill the Internet

Applying Title II to the Internet will destroy it.

---

[59] Beeeeeep boooop kzxrtvzzz, etc.

[60] In fact, bills were introduced in Congress three times to give the FCC this authority and they were voted down. Then the FCC commissioners got together and decided by themselves that they had the power anyway and didn't need to ask Congress to give it to them. Constitution? What Constitution?

You may think that that's an extreme, the-sky-is-falling viewpoint, but that is wrong. First, Title II regulates the Internet like it's 1995. This legislation does not take into account the massive changes that have occurred on the Internet over the last 20 years. It presumes to use an outdated business model that will cripple innovation.

Second, applying Title II to the Internet breaks common load-balancing algorithms that the entire network uses. It's as if the Department of Transportation mandated that there *must* be a road for every car in America. So I have my road to drive on and you have yours. Neither of us can share a road and neither of us can use each other's roads. Imagine how expensive that would be. That is literally what the FCC is doing to the Internet.

> "Those who think that we should turn management of the Internet's infrastructure over to the government had better dig out their 2400 baud modems. Not long ago, that was the "Internet as we know it." Thank goodness it was allowed to evolve." – Larry Downes

Allowing the FCC to control the Internet will drive up the price of access. It will reduce the number of offerings that ISPs provide. And it opens the door for huge new taxes on Internet usage.

But it gets worse. It doesn't take much imagination to realize that the government will start regulating what we can do on the Internet. It's very probable that the government will take action to keep information off of the Internet that it doesn't like.

Government Internet Censorship

You may scoff at the proposition of the government controlling out Internet content. But that would only indicate that you are unfamiliar with what's been going on in Europe.

European Union and several member nations imposed Title II-style regulations on the Internet. Their access is generally slower and more expensive than what's available in the US. Investors won't put their money into an industry that's so heavily regulated that making a profit is doubtful. So the telecommunications industry does not grow like it has in the US[61]. The companies that build the Internet have a chronic shortage of money for investing in making it better. As a result, it just doesn't get better.

---

[61] Martin H. Thelle and Bruno Basalisco, "Europe's Internet Handcuffs Show US How Not To Regulate", Roll Call, August 18, 2013,
http://www.rollcall.com/news/europes_internet_handcuffs_show_us_how_not_to_regulate_commentary-227090-1.html.

While a slower Internet is bad, government censorship is worse.

For example, France has "inspectors" that prowl the Internet to look for content the government deems unacceptable. Germany makes Internet content providers liable for whatever they or their customers put on the Internet. Other countries impose other methods of regulating Internet communications, and thereby killing free speech[62]. Think it couldn't happen in the US? You're so silly.

Does anyone remember the IRS scandal? You know, that little problem we had with the IRS targeting conservative groups because of their political beliefs?

Many liberals didn't mind that, and they said so loudly. But wait until the next Republican administration is voted into office. What happens when it starts using the IRS to target liberal groups? Then it's not so wonderful, is it?

It's not hard to imagine that a presidential administration that uses the IRS to target its enemies will also use the FCC to muzzle them. Make no mistake about this. If the FCC is allowed to continue to regulate the Internet, you *will* see this kind of political targeting. Thinking it will never happen is just plain naïve.

BACK TO THE START

And so I end this chapter where I started it. Simply put, the government is a tool that is effective for a few things and a few things only. History has shown that when the government is overused and too much power is granted to it, that we tip the scales toward tyranny. When that happens, people suffer-and they may even die.

If we are to maintain a functional society with a functional economy, we must recognize what's driving us toward tyranny. It's our money and banking systems, as we saw in Chapter 2. We must rethink and reinvent our whole economy. If we do, we have a chance at restoring the balance between individual and group rights.

In addition, we must be willing to give up our alliances with our silly, pointless political parties. We gain nothing from them. We need to see

---

[62] Peng Hwa Ang, "How Countries Are Regulating Internet Content", The Internet Society, https://www.isoc.org/inet97/proceedings/B1/B1_3.HTM.

government as the Founders saw it and act accordingly. If we can get the government out of the way of the economy like we got it out of the way of the Internet (for a time), something truly wonderful can be born.

It's all there, waiting for us to reach out and grab it.

4 THE FREE MARKET

To understand the miracle that we can build if we simply want to, we first need to understand the basics of the free market.

But let me take a moment to define my terms. When I say "free market" I am *not* talking about Capitalism as we know it today. What we have today is *not* Capitalism—a point that I'll revisit shortly—and it *certainly* isn't the free market.

WHAT IS THE FREE MARKET?

The free market is a large group of people who exchange goods and services because *they choose to*, not because anyone *compels* them to.

While this idea sounds very simple, it actually results in a very complex, self-healing economic network that brings prosperity to those who participate in it. It raises the standard of living for all and provides everyone equally with the ability to rise up to higher levels of wealth.

To understand this idea, let's take a closer look at how the free market works and why it's absolutely awesome.

THE POWER OF THE MIND

The free market is based on the power of your mind. In the free market, you have the power to turn your ideas into a pathway to prosperity for yourself and your posterity. In the process, you improve the lives of those around you.

When people stopped being wandering nomads that followed wild herds, they settled down and planted crops. Those first farmers used the power of their minds to improve their lives. Specifically, they figured out what to plant, how to plant it, how to harvest it, and how to store it.

> "Exploration is the engine that drives innovation. Innovation drives economic growth. So let's all go exploring." - Edith Widder

Next, early farmers used their brainpower to develop animal husbandry so that they could keep livestock like cows, horses, chickens, and so forth.

This process is no different than what happens today. An iPhone programmer, using the power of his mind, came up with an idea for a game that he called Flappy Birds. And that little game became a pathway to prosperity for him.

Now Add Labor and Capital

So just thinking up an idea for a hot, new product isn't enough. You have to do something about it. It doesn't matter whether you're the first nomad to settle down and take up farming or an iPhone programmer creating the latest killer app. You need to make your product real by adding your own labor and some capital.

What is capital? Simply put, it's money or other resources.

"The duty of government is to leave commerce to its own capital and credit as well as all other branches of business, protecting all in their legal pursuits, granting exclusive privileges to none."
- Andrew Jackson

If you're the nomad who wants to be a farmer, your capital may be wild animals that you've captured and domesticated. Or it may logs that you've cut down from the local forest to build your cabin.

If instead you are the iPhone programmer, your capital is money. You use it to buy an iPhone. You need to buy a Macintosh computer to write your iPhone program on. You have to pay a yearly fee to sign up for the Apple Developer Program. Your capital is the money you have to get you started.

In our modern society, people get capital through saving or through borrowing. But borrowing is also savings. Whether you borrow from a bank or from your Aunt Millie, someone else must save up money first in order for you to borrow it. So in reality, all capital comes from savings.

Reaping the Rewards

In a free market, as on a farm, you reap what you sow. That is, if you come up with a great idea for a product, you get the capital you need to create it, you put in the effort to gain the skills to make it, and you get out there and market your product, you can become the next Bill Gates or Steve Jobs.

The free market provides upward mobility. In a free market, you are not locked into poverty because of the circumstances of your birth. You have the ability to lift yourself and your posterity to a higher standard of living. That is the reward of the free market. With imagination, education, and hard work, you can make the same fortune as anyone around you. The free market is an expression of equality that can't be achieved through any other economic system.

Take Steve Jobs as an example. He was the adopted son of a middle-class family. He and his friend, Steve Wozniak, figured out how to build a small computer and sell it. At the time, there were similar computers that you could buy as kits and assemble yourself. But not many people knew how to do that.

The two Steves had just graduated high school. During high school, Steve Wozniak had taken an electronics course and learned how to build computers from scratch. So Wozniak sold his prized HP calculator for $250 and Jobs sold his VW van for $1,500. With that starting capital, Wozniak and Jobs built 10 Apple computers, took them to an electronics hobby shop, and got the owner to sell them on consignment. The computers sole out in a week. Jobs and Wozniak went on to become billionaires[63].

INVESTING FOR TOMORROW

When they started out in business, neither Steve Jobs nor Steve Wozniak knew how to run a business. To get their business to grow, they had to learn a wide variety of business skills and learn them fast.

Learning new skills is an investment that all entrepreneurs must make or they will go under. Likewise, when you are running a business you must reinvest some of your profits into your next product or the market will leave you behind. Investing for tomorrow is part of success.

> "If you want to be truly successful invest in yourself to get the knowledge you need to find your unique factor. When you find it and focus on it and persevere your success will blossom." - Sydney Madwed

When companies invest, they are creating the jobs that will be available tomorrow. If the government steps in and takes away large portions of their profits, then that company's investment in tomorrow is essentially stolen from them. The result of this theft is a host of jobs that are never created.

---

[63]Tim Parker, "Steve Jobs: Legacy of a Tech Guru", Investopedia, September 23, 2011, http://www.investopedia.com/financial-edge/0911/how-much-would-steve-jobs-be-worth-today.aspx.

The more money the government takes, the less money that is available for companies to reinvent themselves into something better. For this reason, a small, limited government is essential to a truly free market.

When you get an education or enroll in an apprenticeship program, you are investing in your own tomorrow. A truly free market will not put up barriers to that investment.

This is an important point. You and I always live between two economies—yesterday's and tomorrow's. Yesterday's economy is the stuff we already have and the stuff we already produce. So anything we're already manufacturing is part of yesterday's economy.

Tomorrow's economy is created by the investments we make in creating new things and offering new services.

The difference between yesterday's economy and tomorrow's economy is important. Yesterday's economy has a limit to the number of jobs it can create. Tomorrow's economy is what creates new jobs, and even entirely new industries.

The more government interferes with creating tomorrow's economy, the fewer jobs we will have in the future and the less our standard of living will grow. It's just that simple.

COMPETITION IS GREAT

The free market rewards people for investing in tomorrow's economy and producing better products that raise the standard of living for everyone. Take the guy who wrote the Flappy Birds game as an example. He made scads of money with his addictive little app. Of course, other programmers see his success and decide they want a piece of that pie. So they learn to write iPhone programs and do their best to come up with a better game than Flappy Birds. If any succeed, it is they who receive scads of money rather than the Flappy Birds guy. If Mr. Flappy Birds wants to keep up his cash flow, he has to come up with a better game than Flappy Birds, or at least a better version of Flappy Birds.

The point here is that competition makes products better. This happens across the entire economy. For instance, there was a time when people that had clogged arteries in their hard *had* to undergo open heart bypass surgery.

This was expensive because very few doctors could perform it. And it was dangerous to the patient.

Then someone came up with a better idea. They wondered if they could smoosh the clogs out of people's heart arteries with a balloon. The did the research, they got the funding, the found that it worked really well, and they made machines and tools for performing the operation. They called this procedure an angioplasty.

The invention of angioplasty meant that fewer people had to undergo risky open heart surgery. It also meant that more doctors could fix clogged arteries. It's far easier to learn to shove the angioplasty tool up into someone's arteries and

> "I have been up against tough competition all my life. I wouldn't know how to get along without it." - Walt Disney

inflate it than it is to learn open heart surgery. Because more doctors could fix clogged arteries, the price of getting them fixed came down[64]. Each doctor who could do angioplasty was now in competition for patients with the heart surgeons—and with each other. One of the best ways to get people to buy your service is to perform it for a lower price. And that is exactly what happened.

Although this is a simplified overview, it show exactly how competition works to drive down prices and improve products and services. The free market improves the standard of living of all who participate in it. The competition inherent in the free market means a greater diversity of products are offered. It also causes the price of products to come down over time. We owe our advanced technical civilization to the operation of the free market.

DISTRIBUTED, DECENTRALIZED, SELF-ORGANIZING

Another great thing about the free market is that no one is in charge of it.

Why is that good?

If someone tried to control the free market, it wouldn't function properly. In fact, the more control you exert over it the worse things get.

In this respect, the free market follows the same principles as the Internet. Right now, there's very little control over the Internet. Anyone can put up

---

[64] Of course, there were also other expenses saved by undergoing angioplasty rather than bypass surgery. They too contributed to the price drop in getting your arteries unclogged. But I've glossed over them for brevity.

information by creating their own web site. Or they can offer products and services. It's up to us what we want to add to the Internet.

If people like what we put on the Internet, our site will get a lot of page visits. If not, no one will look at it. If our product or service on the Internet is useful, valuable, and popular, we can make money with it. If people find it awkward, slow, hard to use, or just plain pointless, we won't be rewarded very well.

The free market works on exactly the same principles. Both the free market and the Internet are distributed, decentralized, self-organizing systems (DDSOS).

We see DDSOSs everywhere around us. For example, a flock of birds is a DDSOS. No one makes any of the birds join up. They enter the flock for their own benefit. As birds migrate, they fly across the sky in the V-shaped pattern that we're all so familiar with. They do this because the aerodynamics of flying in the pattern make it easier for them to keep up their strength while flying. The lead bird, the one at the point of the V, has the hardest time. So all the birds in the flock take turns being the lead bird. They rotate in and out of that role. Again, no one makes them do this. They just do it because it benefits them to take their turn as the lead bird.

"Linux is a complex example of the wisdom of crowds. It's a good example in the sense that it shows you can set people to work in a decentralized way - that is, without anyone really directing their efforts in a particular direction - and still trust that they're going to come up with good answers."
- James Surowiecki

Schools of fish and migrating herds of antelope are other examples of DDSOSs. We see DDSOSs everywhere in nature. But we also see them in human behavior as well. One good example is our system of roads and traffic. No one makes you go out each day and drive around. You do it because it benefits you.

In some countries, people drive on the right side of the road. In others, it's the left side. Either way, it doesn't matter as long as everyone agrees to drive on the same side. And that's the point. There are comparatively few rules when it comes to traffic. And there is very little enforcement. But by and large people obey the traffic rules because it's in their best interest to do so. Within a 50 mile radius of where I live, there are probably 10 million people heading out to work in their cars every morning. All of the cities in that 50 mile circle probably field no more than 10,000 police officers each day to enforce the traffic rules—and it's probably much less than that. But as I said, most people obey most of the traffic laws. Yes, people routinely speed, but if someone goes so fast that they become a danger to everyone else, their likely to be reported to the police.

The point here is that each day an extremely complex system is born, operates, and extinguishes itself with just a few basic rules and very little enforcement. That is how all DDSOSs work, including the free market. Just as the traffic system gives you the freedom to go where you want to go when you want to go there, so the free market gives you the freedom to buy, sell, do business, and make your life better in the way you think is best.

THE SPECIALIZATION OF LABOR

One of the things that drives the free market and makes it produce better products and services over time is the specialization of labor. In the days when the first nomads became farmers, they did everything. They built their own houses, they made their own tools, the produced their own clothing, and so on.

As time went on, the guy who had a lot of cattle and was producing a lot of milk found that it was easier to give some milk and cheese to a guy in the village who was good at making shoes than to make the shoes himself. Likewise, the guy who had a talent for making shoes found it was easier to make shoes and trade them for milk and cheese with the dairy farmer.

Over the centuries, the labor force has increasingly specialized to produce people who have very advanced knowledge in very narrow fields. So, for instance, a guy who helps write the operating system for your computer may know nothing about how his car works. And the mechanic that he takes his car to may know nothing about how to write an operating system for her computer. But they both do their jobs in their area of specialization and it all works out.

Why does it work out?

Money.

In the old days, the dairy farmer and the shoemaker both had to want what the other produced in order to make a trade. In a barter transaction, both parties are both buyers and sellers. Sometimes that didn't work out. There might be instances where the dairy farmer was willing to trade milk and cheese for shoes, but the shoemaker wanted wheat instead. So they were unable to make a deal.

Money allows the specialization of labor and the advancement of technology by decoupling the two halves of a barter transaction. If the dairy farmer sells his milk and cheese to everyone in the village in exchange for money, he can go to the shoemaker for shoes any time he wants. He simply pays with money.

> "Rising living standards - whether in a village, a region, a nation, or the world - depend first on specialization: on letting people concentrate on what they do best and trade with others who specialize in other things." - Virginia Postrel

By disconnecting buying from selling, money enables you to put value into the free market for as long as people will buy your products or services. You can then use the money to take that amount of value back out of the market. And you can take it out in any form you want. You might use the money for shoes, clothes, or to buy a house. Or you might take your family on a vacation to Hawaii. Whatever you find valuable, you are free to buy.

This ability to sell value into the marketplace and later take it back out in any form you want is what enables everyone to specialize in whatever they like doing. History has shown that in countries where there is no money and everyone lives by barter, people don't specialize very much and the local technology won't advance. It's only through the use of money in the free market that we get the labor specialization that results in advancing technology.

THE FREE MARKET IS FREEDOM

In a free market, everyone can offer any product or service that they think will sell. Some will succeed. Others will fail. People vote for your product or service with their wallets. If they like what you're offering, they'll vote for you by paying you. If they don't, then you're out of business. In other words, the free market is the basis of an economic democracy.

The free market is people freely buying and selling. It is people choosing for themselves how to spend their money without being forced. They can choose for themselves using their own judgment, wisdom, and knowledge.

Simply put, the free market is freedom. **People who are against the free market are against freedom.**

THE FREE MARKET REQUIRES CONSTITUTIONAL FREEDOMS

In order for the free market to function properly, it must have the freedoms guaranteed in the US Constitution. Constitutional freedoms enable the free market to exist and provide a means for it to flourish.

Consider freedom of speech. The free market cannot survive without it.

Why? Because innovators are people who challenge the status quo. They are people who speak truth to power.

Take the company Uber as an example. Uber is a new idea in transportation. It matches unmet demand with excess capacity. Specifically, it provides anyone who can pass through a standard vetting process with the ability to give rides to anyone who needs one in exchange for a fee.

So what? Taxicabs already do that.

True, but taxicab companies and drivers are an entrenched profession. That means there are regulations to deal with. And drivers who have to have special licenses. And unions who want a piece of the action.

> "Underlying most arguments against the free market is a lack of belief in freedom itself." - Milton Friedman

Uber sidesteps all of that. You can call an Uber driver with any kind of car you want. If you're looking for a ride for you, your band members, and their instruments, you can be met by an Uber driver with a van. Or if you're going to the airport at a time when there are a lot of other people going there, say at the start of a three-day holiday, chances are you can split the fare with other strangers and all go to the airport at the same time.

The point is that Uber can provide a more customized transportation experience at a much cheaper price. Passengers and drivers can rate each other. So if a passenger is always demanding and complaining, the drivers will avoid him. Or if a driver has a ratty, smelly car, the passengers will choose someone else. For both passengers and drivers, this is an easy and flexible way to either get a ride or make money.

But Uber is new. It challenges the status quo. It tells an entrenched profession, "You've added all kinds of unnecessary, expensive, and anticompetitive barriers to the job of giving rides to people." That's not a popular message in some sectors of the economy.

In a country where everyone must be connected to a political power broker to succeed on any large scale, you don't dare send a message like that to entrenched economic interests. For example, in most of mainland China, Uber could not possibly succeed. At least, not without cutting the communist party bosses in on the action.

> "The most important single central fact about a free market is that no exchange takes place unless both parties benefit." - Milton Friedman

Similarly, in Russia, where mobsters are in control of the government, if you tried to start a business like Uber, you might suddenly find yourself with a "partner" that you didn't know you had. Your "partner" would want a slice of the profits in exchange for "protecting" you.

The lack of due process, protection under the law, limited government, and the freedom of expression is exactly why innovators and entrepreneurs have such a hard time finding success in places like Russia and China. To succeed, they have to curry the favor of those in power.

To extent that the US enforces its Constitutional freedoms, it provides a fertile planting ground for innovation and entrepreneurialism. And when those are lost, so is the free market.

THE FREE MARKET DIED A LONG TIME AGO

Americans like to say that we live in a free market.

Not so much.

We really live in a market that's freer than most. But that's not saying much. The reality is that the free market died a long time ago and what we have now makes us less prosperous than we could otherwise be.

CRONYISM

Whenever a person, a trade association, or a business tries to weight the free market in their favor, they are subverting the market and attacking democracy. While this may seem like an extreme statement, if you stop and think about it that's exactly what's happening.

People looking to cheat the market invariably use the government to do so. To avoid bribery laws, they use various methods to gain favors from government officials. For example, they may promise to bring a lot of jobs into a particular area by establishing a factory or offices there. In exchange, they ask for special tax breaks that others don't get.

By receiving special favors from the government, companies, labor unions, and other organizations make themselves more privileged before the law. In a democracy, we are all supposed to be equal before the law. No one should be able to get favors that aren't available to everyone. Obtaining special favors from the government is actually an act of suborning democracy. And government officials are often all too eager to help. They think that by protecting industries or bringing in jobs they are helping their constituents or gaining votes. But in reality, they are harming small businesses and startups in their areas by making it harder for them to compete. The result is that there ends up actually being *fewer* jobs in their area.

> "The kind of capitalism I hate most is crony capitalism, the friends who decide. These are things which should be killed in Russia." - Anatoly Chubais

Likewise, getting special favors from the government in the form of protectionist legislation that prevents competition destroys any chance that anyone else in the market can come up with a better product or service in that industry.

The telecommunications companies are a perfect example of how favoritism, government control, and protectionist legislation hurt the advancement of the market and cheat consumers. Telecommunications companies by and large owe their existence to the fact that for decades the government prevented other telecommunications companies from operating in their assigned areas. For decade after decade, they gave poor service and a high cost because there wasn't any alternative for customers in their localities.

> "Cable companies aren't bad because they're parts of unwieldy media conglomerates. They're bad because they're monopolies (even where they are no longer legally exclusive) and because the government policies that made them monopolies rewarded lobbying over customer service." - Virginia Postrel

Then came the advent of cell phones. Because this new communications technology did not require stringing wires everywhere, the government couldn't regulate it out of business under existing laws. As a result, the cell phone companies quickly exploded by offering a better product at a better price. Now the landline-based telephone companies that ruled the roost for nearly a century are struggling to survive. Competition has directly benefited consumers, created new product offerings, and lowered the price. This is the free market in action.

To varying degrees, the market we live with now is one of favoritism and cronyism. Some industries, such as the power and water companies, are so heavily regulated that no significant innovation can occur in them. This is a bit of a spoiler, but later in this book I'll show one way we might be able to introduce a competitive market in the electric industry and make it a truly free market. Until such an innovation occurs, we can't possibly have a green, renewable power industry. It's not possible with existing laws. The status quo that is enforced by Big Government and Big Business may prevent us from doing *anything* about global warming. Such is the product of cronyism.

> "Big-government economics breeds crony capitalism. It's corrupt, anything but neutral, and a barrier to broad participation in prosperity." - Paul Ryan

Back in Chapter 2, I showed how the banking industry concentrates power and money to itself. This is the epitome of cronyism and the antithesis of the free market. It was achieved by an unholy alliance between the government and the big bankers through the creation of the Federal Reserve Bank. It will only be when such alliances are illegal that we can have truly free markets.

MYTH: THE GOLDEN AGE OF LAISSEZ-FAIRE CAPITALISM

It's often the case that people use the 19th Century as an example of what they call laissez-faire capitalism, or the free market. They say that the 19th Century was a time when capitalists could do almost anything they wanted. It resulted in exploited workers and huge monopolies.

> "Manufacturing and commercial monopolies owe their origin not to a tendency imminent in a capitalist economy but to governmental interventionist policy directed against free trade and laissez faire." - Ludwig von Mises

Wrong. The 19th Century was the birth of the age of cronyism in America. The free market suffered terribly during that time. Most of the monopolies that were gained during that period were as a direct result of government favors or government interference in the market.

Take the transportation monopolies as an example. The government decided it wanted a transcontinental railroad. Therefore, it granted land to the railroads to build them.

But wait, a portion of that land was already occupied by farmers. No problem. Just seize their land, pay them a nominal fee and call it "fair market value" and give the land to the railroads[65]. Then forbid other railroads from building lines and spurs into the same areas as the existing railroads so that the existing railroads have no competition. And that's how monopolies are born.

I remember as a boy talking with an elderly member of my church congregation. He told a story that his grandfather had told him. It was from this so-called "golden age of laissez-fair capitalism."

This man's grandfather had once owned a farm on the American prairie somewhere in the Midwest. The railroad had come in and would stop at certain points along the track. All the farmers in the area would bring their food and livestock to sell to the railroad men, who would squeeze the farmers for everything they could. The goods and livestock would be bought by the railroaders at prices that could only be considered criminal and then transported to the big cities where they would be sold for a huge profit.

> "We must not tolerate oppressive government or industrial oligarchy in the form of monopolies and cartels." - Henry A. Wallace

The only reason that the farmers put up with this was because they had no choice. Because the government had forbidden any other railroad from doing business in the area, there was no competition. Therefore, the railroaders could pay whatever they wanted.

This member of my church told me and some other youths that were listening to his story that eventually his grandfather couldn't make enough money to stay in business. He left nearly everything behind, took his family, and migrated to the Pacific Northwest where he eventually was able to make a decent living as a farmer.

> "What we do today has nothing to do with capitalism or socialism. It is a crony type of system that transfers money to the coffers of bureaucrats." - Nassim Nicholas Taleb

And the only reason the grandfather found success in the Northwest was because there wasn't a transportation monopoly. The railroad had numerous competitors in the form of boat owners who could ship produce and livestock down the coast. Since both the boats and the railroads were competing for the farmers' business, the prices were better for the farmers.

The point here is that what happened in the 19th Century wasn't a free market. All too often, cronyism between corrupt businessmen and corrupt

---

[65] And even this doesn't take into account that the government had *already* seized the land once from the Native Americans.

government officials resulted in monopolies that enabled the exploitation of others in the economy.

It is true that markets in the 19th Century were in some ways freer than they are today. But that doesn't belie the fact that they were not truly free markets. Free markets require competition. And there are plenty of examples of how the government intervened throughout the 19th Century to create the very monopolies that capitalism haters use as an argument against the free market. This makes their arguments are self-contradictory.

> "In India, innocent and poor children are victims of child labor." - Malala Yousafzai

For example, those who hate the free market often point to the 19th Century and accuse the market of encouraging child labor. Just the opposite is true. Poverty drives child labor.

Prior to the industrial revolution, families had lots of children. There were multiple reasons for this. First, children often did not live to adulthood. So parents had to have enough children so that at least a few of them would live long enough to take care of the parents in their old age. It was a form of retirement insurance.

Second, large families were needed to provide labor on farms, where most people lived. Kids often started farm chores when they were as young as 3 years old. As the industrial revolution progressed, change was slow. It is true that children did work in factories at very young ages. But this was not because of the free market. It was because children were expected to work. If they weren't working in the factory, they worked on a farm.

The free market changes that. In countries where markets are allowed to be free, the specialization of labor that's part and parcel in the free market eliminates child labor. That's because children quickly become unqualified for the available jobs. Children can only provide unskilled labor. But as specialization increases, the need for unskilled labor decreases. People performing jobs need better and better educations or trade skills in order to compete. So children are locked out of the market in a few generations.

Also, as free markets operate in nations, they raise the standards of living. More children live to adulthood. So parents produce fewer children. Poverty decreases and the population stabilizes. This is exactly what happened as we moved from the 19th to the 20th Century. It's still happening in developing nations.

The point here is that the free market does not create the kinds of abuses that were seen in the 19th Century. Most of them had their origins either in cronyism, poverty, or in the customs and standards that had been common for centuries. As the free market operated, it lifted Americans out of poverty and eliminated the very abuses that it's so often accused of.

THE FREE MARKET AND THE GOVERNMENT

Unfortunately, the government thinks that it can manage the free market and "fix" the apparent "defects" in it. Let's examine why this is not so.

REGULATORY BURDENS

In 2014, the Obama administration finalized more than 3,600 new regulations and proposed 2,300 more. The current total cost of federal regulations alone are about $2 trillion (yes, trillion) dollars a year[66].

As we continue to stagnate with record numbers of people not participating in the workforce[67], fully $2 trillion of the nation's GDP is going to nothing more than satisfying government regulators. To do so, they must generate more than 9 billion hours of compliance paperwork[68]. This is 9 billion hours and $2 trillion that is *not* being invested in new businesses–tomorrow's economy. It is not being invested in improving products.

The massive burden of government regulations is also not creating jobs in the marketplace. It's true that the government does employ regulators. But sadly, this does not create jobs. Recall from Chapter 1 that adding government employees actually decreases the

> "Increased spending, growing government debt and overreaching regulations are stifling job creation and economic growth." - Joe Craft

---

[66] Wayne Crews, "CEI Agenda for Congress - Reforming Regulations and Agency Oversight", Competitive Enterprise Institute, https://cei.org/sites/default/files/CEI%20Agenda%20for%20Congress%20-%20Reforming%20Regulations%20and%20Agency%20Oversight.pdf.

[67] There are more than 93 million adult Americans who have dropped out of the workforce, mostly because of the lack of jobs. See Joseph Curl, "Record: Americans not in Labor Force – 93,194,000", The Washington Times, May 8, 2015, http://www.washingtontimes.com/news/2015/may/8/record-americans-not-labor-force-93194000/.

[68] Wayne Crews, "CEI Agenda for Congress - Reforming Regulations and Agency Oversight", Competitive Enterprise Institute, https://cei.org/sites/default/files/CEI%20Agenda%20for%20Congress%20-%20Reforming%20Regulations%20and%20Agency%20Oversight.pdf.

total number of jobs in the economy. This is because the government doesn't spend money as efficiently as private businesses do. The government also pays no penalty for wasting money. So its inefficient use of its funds actually decreases the effectiveness of the economy. Even if an employee in the private sector is getting paid exactly the same as a government employee, the employee in the private sector generates far more value for the economy than the government employee does. Private sector employees simply produce more. Money spent on government employees *always* decreases the total number of employees in the economy.

So even though the government may employ a veritable army of regulators, the salaries paid to these functionaries are a net drain on the economy. And of course, all the money spent satisfying regulators does not go to create jobs in the economy that produce any value.

Yes, people in private industries are paid salaries to produce the mountains of compliance documentation they are required to generate. But this is money that is not invested in production, research, and innovation. And in the end, it is only through production, research, and innovation that we can generate new value. So new stores are not opened. New product lines are not developed. New startup companies do not get much-needed investment cash. New industries are not created. In short, the whole economy is slower, less efficient, and less healthy because of the gargantuan burden of government regulations.

Regulations Do not make Us Safer

But don't we need regulations to make us safer? For instance, what would happen if the Food and Drug Administration (FDA) didn't regulate drugs put out by greedy pharmaceutical companies?

Nothing, that's what.

If the FDA went away, it would not affect our lives at all. Is anyone really so naïve that they think that pharmaceutical companies will take risks by releasing drugs that they know will harm patients? While there have been drugs released that did cause patients harm, Big Pharma isn't stupid enough to do that knowingly despite the plots of many Hollywood movies and TV shows. Big Pharma understands more than anyone how fast the lawsuits will fly when their drugs harm people. They simply won't take chances. They stay in business by bringing drugs to market that help people. They can go out of business fast by hurting them.

Let's be realistic here. Government regulators aren't smart enough to spot problems with drugs and prevent them. Talented people in the pharmaceutical industry don't go into government.

Why?

Because Big Pharma pays more. So who *does* end up in the FDA? Well, it's mostly people who *aren't* that talented. Government regulators tend to be people of mediocre ability who can't get jobs in the industries the regulate. As a result, they aren't going to spot problems that the pharmaceutical companies have missed.

> "There are over 170,000 pages of regulations in Washington, D.C. I want to streamline the rules in the federal government to basically allow businesses to grow without fear of burdensome federal regulations. That's a passion to me, regulatory reform." - Jason T. Smith

Let's ask ourselves, have FDA regulators *ever* spotted and stopped a disaster that the pharmaceutical companies missed? Nope. Not one. This applies across all industries. Where were the government regulators that should have stopped the Great Recession from happening? The financial industry in general and the banking industry in particular are among the most heavily regulated industries in existence. Why didn't *any* of the many regulators in these industries spot what was happening with subprime mortgages and stop the meltdown before it started?

Airplanes have crashed due to mechanical faults. But the endless number of regulators in the FAA haven't prevented that. And when planes do crash due to the fault of the airlines, there are a veritable flood of lawyers waiting with baited breath for the chance to get a grieving family as a client and sue the airlines for everything they've got. And the airlines know this. So

> "The banking collapse was caused, more than anything, by bad government policy and the total failure of bad regulation, rather than by greed." - Nigel Farage

they already do as much as is humanly possible to keep their planes safe. When they do fail, it's a failure that neither they nor the FAA regulators could possibly have spotted no matter how many regulations were on the books.

Regulations and regulators do not make us safer. The FDA doesn't really do anything at all. Neither does the FAA, or the SEC, or any of the other host of government regulators. If they all disappeared tomorrow, there would be few if any negative effects. And the positive effect would be that an extra $2 trillion would go into the economy to create jobs, new products, new businesses, and new levels of prosperity.

If you really believe that government regulators make us safer, then you must not know about the revolving doors that exist between the government agencies and the industries they regulate.

> "'Security theater' refers to security measures that make people feel more secure without doing anything to actually improve their security. ... Our current response to terrorism is a form of 'magical thinking.' It relies on the idea that we can somehow make ourselves safer by protecting against what the terrorists happened to do last time." " - Geoff Davis

For example, talented government regulators in the Securities and Exchange Commission often find jobs with the very companies they oversee[69]. Regulators looking for positions often are quite sympathetic to the interests of the companies that might hire them[70]. In fact, it's so common for regulatory agencies to become dominated by the companies that they're intended to regulate, that economists and political scientists have coined the term *regulatory capture* to describe the phenomenon[71].

Does anyone think that an FDA that is largely governed by Big Pharma is keeping us safer? Is the SEC, with its incestuous relationships with High Finance going to actually protect us? Not so much. An Environmental Protection Agency (EPA) that constantly makes nice with Monsanto Corporation, the largest pesticide producer in the world, is not really protecting anyone from pesticide pollution.

> "I stood there, an American citizen, a mom traveling with a baby with special needs formula, sexually assaulted by a government official. I began shaking and felt completely violated, abused and assaulted by the TSA agent. I shook for several hours, and woke up the next day shaking." – Erin Chase

Likewise, the Transportation Security Administration, doesn't stop terrorists from killing us on airplanes. It's all just for show[72]. It's theater designed to make people *feel* safer[73]. You're not getting any added protection from all those TSA regulations that prevent you from carrying a water bottle onto a plane. In fact, TSA agents fail 96% of the tests that the government performs so see whether the

---

[69] "Dangerous Liaisons: Revolving Door at SEC Creates Risk of Regulatory Capture", The Project on Government Oversight, 2013, pg 3-4, http://pogoarchives.org/ebooks/20130211-dangerous-liaisons-sec-revolving-door.pdf.

[70] Ibid. pg 5.

[71] "Regulatory Capture", Investopedia.com, http://www.investopedia.com/terms/r/regulatory-capture.asp.

[72] The TSA's own documents show that it doesn't think terrorists are going to try another attack on airplanes anyway. See http://www.scribd.com/doc/178449913/Corbett-Files-2-unredacted.

[73] "Expert: TSA Screening Is Security Theater", CBS News, July 31, 2009, http://www.cbsnews.com/news/expert-tsa-screening-is-security-theater/

agencies will detect and stop people from carrying bombs and weapons onto planes[74].

To add insult to injury, TSA employees routinely steal from you[75]. And to be brutally honest, they use the so called "naked scanners" to ridicule and ogle at travelers[76]. The idea that these regulators are helping you in some way is sadly and laughably false.

Regulating Businesses Equals Nationalizing Them

Most people know that nationalizing a business is when the government takes both ownership and control. This is not uncommon in Marxist countries. It is a hated symbol of government oppression in the US.

But did you know that if you regulate a business enough, it's really the same as nationalizing it?

It's true. Government regulations can so completely tie the hands of a business that it is the same effect as nationalizing the business outright. At some point, we have to admit that varying degrees of Marxism are creeping into our economy by the slow poison of government regulation. By slowly increasing government regulations over time, one industry after another falls under control of the government.

This seems paradoxical given the fact that big industry players often do their best to dominate government agencies through regulatory capture. What we really have is a pointless, unnecessary, and expensive war between the government and the companies that it regulates. In some industries, the companies dominate and take over the government agencies. In others, the government dominates and indirectly nationalizes the companies under its control. Either way, we lose.

---

[74]"Investigation: Undercover agents snuck fake explosives, banned weapons past TSA", Fox News, June 1, 2015, http://www.foxnews.com/politics/2015/06/01/investigation-undercover-agents-snuck-fake-explosives-through-tsa-checkpoints/

[75] See "TSA Agent Angel Velazquez Accused Of Stealing $8,500 From Checked Luggage", Huffington Post, January 31, 2014, http://www.huffingtonpost.com/2014/01/31/angel-velazquez-tsa_n_4703136.html and "Ex-TSA agent: We steal from travelers all the time", RT.com, September 30, 2012, http://rt.com/usa/tsa-stealing-from-travelers-358/.

[76] Jason Edward Harrington, "Dear America, I Saw You Naked and Yes, We were Laughing. Confessions of an Ex-TSA Agent", Politico Magazine, January 30, 2014, http://www.politico.com/magazine/story/2014/01/tsa-screener-confession-102912.html#.UuvBTnddVk4

Regulating Businesses Destroys Innovation

When businesses are subject to regulations, the regulators often take a "guilty until proven innocent" stance. Business owners must go out of their way to show that they are not doing anything on the ever-increasing list of things that regulators say is forbidden. In fact, it's getting so bad that sources say that "small business owners [report] that with many inspections, both federal and local, it's a matter of negotiating what you're going to be cited for, not trying to be in full compliance, since there's no way to actually comply with this massive tangle of regulations."[77]

> "People should be free, people should be unencumbered by regulation as much as possible, that big government always goes corrupt and the truth shall always set you free." - Glenn Beck

According to a Gallup poll, 22% of small business owners say that their most important problem is "complying with government regulations."[78] Because the vast resources that businesses must expend on complying with government regulations, time and money isn't being spent on developing the Next Big Thing. At least, not nearly as much of it isn't being spent as could be.

Regulations freeze the market as it is. They assume that yesterday's economy will continue into the future indefinitely–that tomorrow's economy will never be invented.

For instance, regulations that protect taxicab drivers assume that a company like Uber will never be created. They often prevent Uber from doing business in particular cities. They take yesterdays' business model and freeze it in place for all time. This makes it impossible for someone to come along with new business models. It also makes it impossible for innovators to increase the number of jobs in the economy.

Let's compare the banking industry with the high tech industry as an example. In the late 1990s, Steve Jobs had Apple build a new MP3 player called an iPod. Other than the cool name and the wheel interface (which Apple has since abandoned), there wasn't that much new about the iPod.

On the other hand, the iPod's business model was revolutionary. Back then, most makers of MP3 players expected you to buy records (remember records?), tapes (also gone now), or CDs (also nearly gone). You could then

---

[77] Megan McArdle, "Federal Regulations Kill Innovation Too", The Daily Beast, February 5, 2013, http://www.thedailybeast.com/articles/2013/02/05/federal-regulations-kill-innovation-too.html.
[78] Adrian Moore, "Don't Be Fooled, Regulations Cost Jobs", Real Clear Markets, August 9, 2012, http://www.realclearmarkets.com/articles/2012/08/09/dont_be_fooled_regulations_cost_jobs_99812.html.

use one method or another to copy your music onto your computer. From there, you could copy it to your MP3 player and listen to it.

It's a wonder any MP3 player manufacturers could do business at all.

But the erstwhile Mr. Jobs invented a completely new business model. He decided to package an MP3 player with an online store. You could buy your music *right from your computer!* Not only that, it was already in a format that was ready to be played on your nifty new iPod. So in minutes, you could buy music, download it, and put it on your iPod. This was heady stuff.

Jobs further revolutionized his business model by later making it possible to buy music *right on the device itself*—no computer needed. You bought your music over that newfangled Internet thingy that everyone was talking so much about. Consumers worldwide screamed, "Shut up and take my money!" Steve Jobs had created a hit.

Imagine for a moment what would have happened to Steve Jobs if he had faced the kind of regulations that are common in the banking industry. There are so many barriers to entry in the banking industry that any new idea takes years and millions of dollars to deploy to the market. So it's almost impossible to try anything new. That's why we use a banking model that has been in place since the 1600s.

The high tech industry is where most of today's innovation is happening. And that's precisely because it's nearly completely unregulated. Both hardware and software manufactures can innovate, bring new products to market, and produce new services with almost no regulatory barriers. This is exactly what gives us companies like Uber, Facebook, and so forth.

> "The media has brainwashed the electorate to expect the government to do something. The best economic policy of any government is to do nothing but reduce the size of the government, reduce the size of the laws, and reduce the size of regulations." - Marc Faber

Can you imagine what it would be like if new social networking companies had to comply with government regulations before they could be deployed? MySpace would still be the Big Thing, perhaps supported by government subsidies. While Facebook, Tumblr, and Vines would never have gotten a chance.

Regulations necessarily assume that nothing in the free market will change. And once they're imposed, nothing *can* change because the changes don't comply with existing regulations. It's really just that simple.

Government Regulations in Short

At this point, I've gone on and on about the negatives of government regulations. But as I said in Chapter 3, the government is a pair of pliers. Pliers should not be used to hammer nails. You shouldn't use them to turn nuts or bolts.

Similarly, there's a lot that the government is just not good at. So we shouldn't impose layer upon layer of silly regulations on those tasks. However, that doesn't mean that we should throw away *all* government regulations. We just have to remember that any government regulations we impose will have a negative impact on the economy in general and on innovation in particular. There may be some things that are worth that negative impact.

> "Subsidies and mandates are just two of the privileges that government can bestow on politically connected friends. Others include grants, loans, tax credits, favorable regulations, bailouts, loan guarantees, targeted tax breaks and no-bid contracts." - Charles Koch

For instance, a friend of mine recently bought a piece of property to build a house. After a little digging, he found that the previous owner never sent his garbage to the dump. He buried it on the property. The whole property had mounds of buried garbage just under the surface. Before he could build his house, my friend had to dig up and haul away that garbage.

A situation like this clearly shows that there are people who are willing to pollute no matter what. It is right that it should be illegal to dump pesticides, old cars still filled with gas and oil, or hazardous chemicals underground or in our water supply. The people responsible for such things should be forced to clean them up.

> "Why has it seemed that the only way to protect the environment is with heavy-handed government regulation?" - Gale Norton

But that's not where we are now. Another friend of mine recently had a bonfire in her back yard for her teenagers and their friends. This was a nice church lady providing good clean fun for some very outstanding young people. What she was doing was perfectly legal according to local laws and EPA regulations.

However, an EPA inspector saw the smoke and barged onto her property to investigate. Note that this EPA inspector had no prior knowledge that a crime was being committed. In legal terms, he had no *probable cause* to come onto her property. If he was a county sheriff, it would have been against the law for him to even show up. But for some reason, the EPA gets a pass when it comes to our Constitutional rights. They can go onto anyone's private property at any time of the day or night with no probable cause and no reason to believe a crime is being committed.

In any case, this little dictator of an EPA inspector invaded my friend's property and saw that she was having a bonfire that completely conformed to all laws and regulations. However, because he needed to get his quota of tickets written, he cited her for starting the fire with newspaper. He told her, "It's illegal to burn garbage."

In spite of my friend's protests that she wasn't burning garbage and that she had just used the newspaper to start the fire, the EPA inspector pompously lectured her and then fined her $2,000. When she said she'd fight it in court, he literally laughed. You can't fight the EPA in court. They're above the court system. In the end, she had to knuckle under and pay the exorbitant and ridiculous fine.

The point here is that the government's regulatory system is not only out of control, extreme, arbitrary, and downright dictatorial, it is unconstitutional. The radical amounts of regulations that businesses have to comply with these days is one of the reasons our economy has stagnated for years. It's just too much and it's time to dial it back to something more sane.

YOU CAN'T SPEND YOUR WAY TO PROSPERITY

After the Great Recession of 2008-2009, we saw politicians and government-backed economists race around and try to get America to start spending like there was no tomorrow. Their idea was that spending would stimulate the economy and get production going again. Jobs would increase, giving more people more money to spend. If we just spend, spend, spend, all our problems will disappear.

Fortunately, the American people didn't listen.

What's wrong with this thinking? Well, it ignores some basic facts about the economy.

The Free Market Stinks When It's Rotten

We need to realize that if there's something wrong in the economy, then the economy is not doing the right thing. If the market stinks, then something's wrong with how it's functioning.

In the case of the Great Recession, government policies distorted the housing market and created an artificial price bubble the finally burst. The distortion of housing prices caused people to invest in the wrong thing. Specifically, they were investing in yesterday's economy—the economy that produced tangible assets like houses. What they are *not* investing in was tomorrow's economy—the economy that produces new jobs, new markets, and greater prosperity.

"What's hurting the US economy is total government spending. The deficit is an indicator that the government is spending so much money that it can't even get around to stealing all of the money that it wants to spend. But the tip of the iceberg is not what hit the Titanic - it was the 90 percent of the iceberg under water." - Grover Norquist

Investors, and regular homeowners, ended up with lots of wealth on paper in the form of high house values. But then they ended up not being able to pay their exorbitant mortgages. So the price bubble burst and erased huge amounts of their wealth. This proves that although spending did stimulate the economy briefly, it created more problems than it solved.

Spending doesn't fix problems in the economy—especially not in the long term. Saving *does* fix economic problems. Why? Because when you save money, your money goes into a bank. The bank doesn't just let it sit there in the vault. They lend it out to people who are starting businesses, investing in upgrades for their companies, and so on. The money gets used for making businesses go, investing in new products, buying houses, college loans, and so forth. In other words, it gets invested in tomorrow's economy and creates more jobs and more tangible goods, as well as a higher standard of living and new opportunities for everyone.

In the Great Recession, the price distortions in the free market were caused by government policies incentivized spending on housing. The money did *not* go into investing in new business ideas, new products, upgrading existing businesses, and innovation. So although there was a brief period where everyone had lots of money on paper, they didn't increase their actual wealth. In fact, their spending *decreased* their wealth.

Innovation Drives the Market

Let's look back to the 1980s and 1990s. It was the beginning of the personal computer revolution. In 1980, almost no one owned a computer. By 2000, almost everyone had a computer on their desk at work, another one in their home, and we were starting to carry them in our pockets.

During that 20 year period, innovation was rampant. And with the advent of the Internet, that has only continued. In the 1980s, both Apple Computer and

Microsoft had a handful of employees each. Today, these are some of the largest employers in the computer industry. What happened? Innovation and the chance to compete in the free market happened all over the place.

> "The heart and soul of the company is creativity and innovation." - Bob Iger

Apple and Microsoft aren't the only ones. Google, Twitter, Facebook, Amazon, Skype, eBay, Pintrest, and a host of others have appeared. These companies not only created jobs directly, most of them have created jobs indirectly. Nearly all of them provide the infrastructure for individuals and small companies to compete as effectively as large companies.

Let's use eBay as an example. Although eBay doesn't have that many employees (competitively speaking), there are a vast host of individuals who earn a living selling things on eBay. In fact, some estimates put the number as high as 1.3 million people[79].

Amazon is the same way. The company may employ tens of thousands of people, but it also enables a veritable army of millions of online sellers to make a living. Likewise, both Apple and Google have created app stores in which small scale developers, often individuals, can write and sell their apps to earn part or all of their living. Does anyone remember Angry Birds? That was just a handful of people who wrote a very addictive game. They were able to go head to head with big game development companies and make tons of money.

Some economists bewail the new Internet-based high tech companies. They say that companies like Amazon are job killers because they put retail stores out of business. Unfortunately, that's 19th Century thinking.

> "Learning and innovation go hand in hand. The arrogance of success is to think that what you did yesterday will be sufficient for tomorrow." - William Pollard

The days of behemoth companies employing hundreds of thousands of workers from cradle to grave are long gone. These days, companies like Apple, eBay, and Amazon let people start their own small companies and go toe to toe with the big fish in the pond. Companies like this are called *aggregators*. They make their money by aggregating together lots of small businesses and providing them with the infrastructure they need to compete at the same level as big companies. That's the new business model in the age of the Internet.

---

[79] Daniel Gross, "My eBay Job", Slate, May 21, 2008,
"http://www.slate.com/articles/business/moneybox/2008/05/my_ebay_job.html"

The highly distributed, fast-paced, innovative market that is a direct result of Internet innovation is far more healthy than the old way of doing things. Small business competing in the free market is the greatest engine of prosperity that the world.

Apple started with Steve Jobs and Steve Wozniak literally working out of a garage. It's because small businesses like that can compete in the free market that we have things like iPads and iPhones. Apple grew from two guys in a garage to one of the most valuable corporations in the world. It's precisely because small businesses can spring up in America when they want to innovate that we saw the prosperity that came in with the computer revolution.

"It isn't the incompetent who destroy an organization. The incompetent never get in a position to destroy it. It is those who achieved something and want to rest upon their achievements who are forever clogging things up." - F. M. Young

The advent of the Internet has only made it easier to create our own jobs by repeating the story of Apple over and over. Although the Internet has destroyed many traditional big-company jobs, it has created far more new-economy jobs than it ever destroyed. It is the greatest innovation tool ever invented in human history so far.

Let me take a minute to make one more mention of the blockchain. The blockchain is the new engine of innovation. It will be at least as impactful as the Internet. And it's entirely possible that when future generations write the history of the 21st Century that they could list the blockchain as the 21st Century's most important invention.

Innovation Requires Investment

The small companies that constantly innovate in the free market don't go anywhere without investment. Without investment, we don't get prosperity. If there is no investment, we don't get cool new gadgets that we just have to have no matter what. And that's the lesson of the Great Recession. If we misspend our money on overpriced houses, we don't get prosperity as a result. We usually end up with less than we started with.

On the other hand, if there is a lot of money available for investments in new, innovative ideas, then the economy prospers. New jobs get created. Companies appear out of nowhere. We've seen it over and over since the start of the computer revolution.

Investment Requires Savings

Where does investment money come from? Savings. In order for anyone to invest, someone must save first. Savings finances innovation. Spending does not. Spending makes the economy do more of what it's doing right now. It doesn't matter whether the economy is doing the right thing or the wrong thing. Whatever is happening, spending will make it do more of the same thing. So spending may help or spending may hurt. But saving always helps.

A healthy economy *requires* that people save. If people are spending everything they have, it's only a matter of time before the economy starts to unravel.

GOVERNMENT SPENDING AND THE FREE MARKET

Sadly, many politicians and economists miss the basic fact that savings creates a healthy free market not spending. For example, the economist Paul Krugman wrote a laughable book called *End This Depression Now!* in which he endlessly lauded the virtues of spending, spending, and more spending.

Krugman was supremely unhappy that consumers wouldn't follow the government's directives to be spendthrifts. So he advocated increased government spending.

Janet Yellen, the current head of the Fed, agrees. She said that by having the Fed take actions to drive down interest rates, that an economic stimulus would result by people spending more. That, in turn, would create more jobs and fix everything[80].

No. Ms. Yellen's approach completely ignores the cost of inflation that's produced through the Fed flooding the market with new cash in the form of Quantitative Easing. It also ignores the fact that all that spending will not fix the underlying problem; the free market was distorted by government intervention. More government intervention cannot repair that. Millions of people saw much of their wealth erased. You can't spend wealth back into existence.

> "The same undisciplined government spending and social engineering that has undermined our economy over the past 30 years has also been tearing at the social fabric of this land." - Stockwell Day

As I've mentioned previously, government spending to create wealth can't work anyway. In the free market, individuals and businesses eventually feel

---

[80] Rana Foroohar, "Janet Yellen: The 16 Trillion Dollar Woman," Time Magazine, January 20, 2014.

the pain of money they misspend. That's why they generally tend to spend wisely. The government never feels that pain. Government workers and officials can waste endless amounts of money and never face any penalties for it. As a result, they never stop wasting money.

Money wasted by the government is money that could have been spent more efficiently by individuals and businesses. Or it could have been saved by individuals and businesses. That savings would have been made available by banks for investments in innovation. Instead, it gets misspent and encourages businesses to do more of the wrong things.

Calls for more government spending never work. In fact, they *can't* work. As Forbes writer Doug Bandow puts it, "If government could spend America to prosperity, good times would have arrived long ago."[81]

Government spending makes the free market suffer. It's far better to leave the money in the hands of individuals and businesses who will spend it more wisely. Or they can save it in a bank where it will end up being available for investments that are proven to drive prosperity.

YOU CAN'T TAX YOUR WAY TO PROSPERITY

Back in 2011, some prominent statist[82] writers published an article that asserted we could jump-start economic growth not only by spending, spending, spending, but by taxing, taxing, and more taxing[83]. Well government has spent and taxed with exactly the vigor these authors advocated. What do we have as a result? More stagnation. Their prescription hasn't worked in the slightest.

Taxes are penalties for success. Whatever you tax decreases. Whatever you fund increases.

We have taxed spending, saving, doing business, and investments. The result? We have less of all of them.

---

[81] Doug Bandow, "Federal Spending: Killing the Economy with Government Stimulus", Forbes, August 6, 2012, http://www.forbes.com/sites/dougbandow/2012/08/06/federal-spending-killing-the-economy-with-government-stimulus/

[82] A statist is someone who worships big government.

[83] Catherine Hollander and Jim Tankersley, "18 Ways to Jump-Start Economic Growth: Spending Money Edition", National Journal, August 11, 2011, http://www.nationaljournal.com/economy/18-ways-to-jump-start-economic-growth-spending-money-edition-20110811

On the other hand, we have funded poverty, non-participation in the workforce, out-of-wedlock births, corporations who do business badly, banks that make foolish loans, farmers who don't farm, and a host of other dysfunctional behaviors. The result? We have far more of all of them than ever before.

> "I am favor of cutting taxes under any circumstances and for any excuse, for any reason, whenever it's possible." - Milton Friedman

Increasing taxes means less investment. Increasing taxes creates a wasteful industry in tax avoidance.

Decreasing taxes always brings about increased economic activity[84]. Cutting taxes results in a healthier economy.

SAVINGS INCREASE INVESTMENT

Saving money takes discipline. People generally don't like that. But saving money means we have a buffer to fall back on when difficult times come. Every American should have a savings account with enough money to cover a year of their expenses without working.

That's a bold statement. But if Americans took it seriously and actually had a year's supply of money in the bank, they could weather life's storms fairly easily.

Not only that, if Americans had a year's salary saved in their bank accounts, the money that they held would spur the economy like almost nothing we've seen in decades.

Remember that money held in savings creates investment capital that can be used to innovate, improve businesses, start new businesses, or come up with that great new killer website. Money that we save spurs investments. It provides a shot in the arm for the economy. And that is always something that's helpful to the free market.

So it's savings that creates prosperity, not regulations, not government spending, and not taxes.

---

[84] Andrew G. Biggs, Matthew H. Jenson, "Yes, You Really Can Cut Your Way to Proserity", American Enterprise Institute, July, 14, 2011, https://www.aei.org/publication/yes-you-really-can-cut-your-way-to-prosperity/.

THE FREE MARKET DEFEATS POVERTY

People who are uninformed about what the free market really is often blame the free market for inequality and poverty. What such people are not seeing is that our current markets are filled with corruption and cronyism. And *that* is exactly what drives poverty.

As we saw in Chapter 2, that government issued debt-based money controlled through fractional reserve banking actually forces people into poverty no matter how well they manage their money and how hard they work. This is not a product of the free market but of cronyism and government corruption. The correct role for the government is not to manage the market, regulate every aspect of it, and grant favors to well-connected special interests. This is corruption pure and simple.

> "Every nation on the Earth that embraces market economics and the free enterprise system is pulling millions of its people out of poverty. The free enterprise system creates prosperity, not denies it." - Marco Rubio

Governments that allow cronyism to flourish invariably lead to a patronage-style system similar to what is common in China, Latin America, and much of the developing world. Unelected government bureaucrats should not have the power to regulate entire industries out of business. Neither should they have the power to grant government-backed loans to special interests such as solar power companies, farmers, and so on.

The free market provides more ways out of poverty than any other system. We can see this just by looking at the differences between more and less developed nations. Severe poverty has disappeared in most industrialized countries. These same countries are the closest to being free market nations. Less developed nations invariably have a culture of corruption and cronyism deeply embedded in both their government and business sectors.

There are nations, such as India and China, that were very Marxist in the past. They were riddled with corruption and cronyism. But as both India and China have moved in the direction of freer markets, poverty has abated on a vast scale. Literally hundreds of millions of people rose out of grinding poverty in India and China over the same time period that they moved closer to free markets[85].

---

[85] D.W. MacKenzie, "The Data is Clear: Free Markets Reduce Poverty", Mises Daily, The Mises Institute, June 16, 2014, https://mises.org/library/data-clear-free-markets-reduce-poverty.

In fact, over a twenty year period, the free market has lifted nearly one billion people worldwide out of extreme poverty. It's true. Between 1990 and 2010, almost one billion people in developing nations rose from horrible, grinding, unfair, inhumane poverty[86]. Before 1990, about 43% of the population in developing nations lived on less than $1.25 a day, which is the international definition of poverty[87]. At that level of subsistence, people's lives are "poor, nasty, brutish, and short."[88]

But free market reforms around the world has changed all that. In just twenty short years, a billion people are living better lives because of freer markets.

I'll say it again, the free market provides more paths out of poverty than any other system. People don't need handouts, social programs, giveaways, and so on. What they needs is a job, a chance at a real education, and the freedom to build whatever business they think will succeed.

We as a society can build an economy that resists cronyism. We can build something that is freer and more humane, and that encourages more innovation, and brings the power that High Finance and Wall Street have horded for themselves down to the level of everyone on Main Street. We'll see how to do that starting in Chapter 6. But first, let's go on to Chapter 5 and take a look at Marxism, which is the biggest competitor to the free market.

> The great virtue of a free market system is that it does not care what color people are; it does not care what their religion is; it only cares whether they can produce something you want to buy. It is the most effective system we have discovered to enable people who hate one another to deal with one another and help one another." - Milton Friedman

---

[86] "Towards the end of poverty: The world's next great leap forward", The Economist, June 1, 2013, http://www.economist.com/news/leaders/21578665-nearly-1-billion-people-have-been-taken-out-extreme-poverty-20-years-world-should-aim.
[87] Ibid.
[88] Ibid.

5 WHAT ABOUT MARXISM?

We saw in Chapter 4 that cronyism has corrupted the free market and is centralizing control over all aspects of human life, starting with the economy. People who are not familiar with what the free market is and who fear freedom can't see the difference between cronyism and the free market. They look to other ways to conduct the economy and government.

All too often, people today are proposing that we dig into the "ash-heap of history"[89] and resurrect the failed philosophy of Marxism. Marxism includes multiple variants, most notably Leninism, communism, and socialism.

The primary idea behind Marxism is that humans are basically animals that were shaped by evolution. Marx thought that if the proper society were invented, it could shape humans into a perfect individual. He wanted to accomplish this by transferring the rights and powers of the individual to the society. This would eliminate those powers and therefore people couldn't take advantage of one another as they so often do.

The goal was to eventually produce a race of perfect people that would not act against each other and live in harmony together. In other words, it was Marx's path to Utopia.

COULD CENTRAL CONTROL WORK?

The entire basis of Marxism depends on total government control of all individuals in society. Without total control, there will always be dissenters and contrarians who don't want to play along. The government *must* have the power to "re-educate" such people, punish them, or execute them if the greater good of a perfect society is to be achieved.

The first thing that Marxists miss with this is that by centralizing power, you also magnify it. And you magnify it a lot[90]. The impacts of the central

---

[89] Ronald Reagan, "Address to Members of Parliament," The Heritage Foundation, June 8, 1982, http://www.heritage.org/research/reports/2002/06/reagans-westminster-speech
[90] Friedrich A. Hayek, "The Road to Serfdom," The Institute of Economic Affairs, 2005, p. 40.

planners reverberates throughout the economy in ways that are seldom achieved in the free market.

To Marxists, the magnification of power is a good thing. It means that their programs are more effective. To the rest of us, it means that the central planners can take total control of our lives and there's really nothing we can do about it.

> "A glance at the economic system and methods of totalitarian states — of the Soviet bloc, for example — is enough to show that state-ownership of the means of production does not lead to an increase of wealth for the people but, on the contrary, to their exploitation, whereas the reverse is true of the free countries and peoples, which are denounced for their so-called capitalism but which clearly illustrates how private ownership of the means of production is contributing more and more to the general welfare." - Ludwig Erhard

It also means that the central control of all wealth. Money, property, and any means of production must be seized by the state and centrally controlled. This wealth is redistributed in ways that the central planners think are fair, advantageous to their goals, or just however they feel like.

Under Marxist theories such as communism and socialism, no one should be allowed to accumulate wealth through their own hard work, education, careful planning, and judicious investing. You have nothing except what the central planners give you.

LIFE UNDER CENTRAL PLANNING

What's it really like to live in a centrally planned society? The best way to find out is to ask people that have done it.

A few years ago I spent some time in the Republic of Kazakhstan. I had a chance to find out firsthand what happened under Soviet rule.

Mine, Mine, All Mine

The Soviets appropriated Kazakhstan's vast natural resources. Kazakhstan has gold, oil, and uranium, among other things. The Soviets "redistributed" this vast wealth in ways that benefited themselves. They left the people of Kazakhstan in poverty.

Kazakhstan was the location of some of the Soviet Union's most infamous gulags. In case you, gentle reader, are too young to remember the days of gulags, a gulag is a labor/prison camp. If you needed "re-education" in the delights of communism, you were sent there to do hard labor and produce goods for the glorious motherland. A quick look at the history of the Soviet Union shows that they performed economic miracles in modernizing and advancing their country. They did it on the backs of the people in their satellite states that they impoverished by "redistributing" their wealth and on the back of literally millions of people in slave labor camps. Let's face it, any country can accomplish economic miracles if they can steal enough resources and enslave enough people.

When the Soviet empire fell, Kazakhstan had very little in the way of industry in spite of its abundant resources. Kazakhstan didn't even have its own form of money. All of the Soviet satellite states were forced to use the Soviet ruble. As the USSR collapsed, the Russians promised the various parts of their former empire that a new version of the ruble would be issued. It would not longer contain the picture of Lenin, and the Russians would exchange the old rubles for new rubles one for one.

> "If you look at great human civilizations, from the Roman Empire to the Soviet Union, you will see that most do not fail simply due to external threats but because of internal weakness, corruption, or a failure to manifest the values and ideals they espouse." - Cory Booker

Ha.

The day came when the promised new rubles were supposed to arrive in Kazakhstan and all of the other former Soviet republics. But there were no rubles. People in the newly-liberated republics had no choice but to cross the border into Russia and exchange their old rubles for new ones. But unfortunately the promised one-for-one exchange rate was nowhere to be found. The citizens of the republics suddenly found themselves with far less money than they thought they had.

Most of the republics went into instant economic collapse. Some of the republics still use Russia's ruble. Others have issued their own currency.

In Kazakhstan, the local leader of the communist party, Nursultan Nazarbayev, became the president of the country more or less by default. He knew the Russians well from his long association with them in the party, and he realized that something fishy was going on with the currency. He decided to take action.

> "Communism is not love. Communism is a hammer which we use to crush the enemy." - Mao Zedong

Instead of waiting for the promised new rubles from Russia, Nazarbayev pulled off one of the most astute pieces of political leadership in modern times. He literally loaded entire cargo jets with gold from Kazakhstan's mines and flew them into Western Europe. Once there, he put the entire gold shipment on deposit at a European bank. He then contracted with a Dutch company to design and produce a new currency called the tenge (pronounced TEN-geh). He then shipped planeloads of the tenge back to Kazakhstan.

> "Frankly speaking, I decided to become a businessman at the moment when I understood that it is possible, because I grew up in a country where it was not possible. There existed even a special article in the penal code of the Soviet Union which punished entrepreneurial activity." - Vladimir Potanin

On the day that all of the other former Soviet republics found out that they'd been had, Nursultan announced his new currency. He told his citizens to go to their local banks to exchange their Soviet rubles for tenge. The tenge was backed by gold, so it had instant value. Nursultan had saved his country from ruin at the hands of the Russians.

The Reality of the Soviet Union

Kazakhstan has two main ethnic groups. The first is the indigenous Kazakhs, who are closely related to their neighbors the Mongols. From what I was told while in Kazakhstan, the Russian Czars killed about half of all living Kazakhs when they conquered the area.

After the fall of the Czars the Soviets came through and killed half of the Kazakhs that remained. Today, the Kazakhs are a minority in their own country.

> "In the 20th century, the Soviet Union made the state's role absolute. In the long run, this made the Soviet economy totally uncompetitive. This lesson cost us dearly. I am sure nobody wants to see it repeated." - Vladimir Putin

There are still monuments in Kazakhstan to the brave horse-mounted warriors that fought for Kazakh independence during that era. They are both inspiring and sad to look at. These mounted nomadic cavalry fighters valued their freedom so fiercely, that they were willing to go up against tanks and motorized, mechanized troops to fight for it.

They stood no chance, and they eventually submitted to Soviet rule.

The second ethnic group in Kazakhstan is the Russians. Some have moved in recently. The ancestors of others came during Soviet times. Some are the descendants of people that were interred in the gulags. Still others are the descendants of Russians that arrived during the time of the Czars.

Russians settled in Kazakhstan largely for farmland. Kazakhstan is a vast, open sea of grass. The Soviet Union undertook a huge resettlement program in which they brought in Russians to till the soil. The Kazakhs were largely nomadic horsemen. It took Soviet military might to settle them down and stop their nomadic wanderings.

For generations, the Russians controlled the country. That is changing now that Nazarbayev is in charge. Nazarbayev, himself a Kazakh, has done a lot to promote the preservation and teaching of the Kazakh language and culture. Kazakhs are sent to different schools than local Russians. They learn Kazakh as their first language, which was discourage under Soviet rule. Both the Kazakhs and the Russians get the same standard educations, but the Kazakhs get theirs from Kazakh teachers teaching in Kazakh, while the Russians are taught by Russian teachers in Russian. I was never able to find out what schools the increasing number of mixed race children go to.

The Largest Environmental Disaster on Planet Earth

The influx of Russians into Kazakhstan and the surrounding Soviet republics and the establishment of their farms directly resulted in the largest environmental disaster that has ever occurred on Earth.

No, I'm not talking about Chernobyl. As bad as that was and still is, Chernobyl is one of the more minor disasters that Soviet central planning and economic control caused.

Kazakhstan, like most of its neighbors, is largely grassland. The Soviets saw the endless grass ocean (larger than the continental US) as a great new source of farmland. But there was a reason that the indigenous people were nomads and not farmers. That reason was water.

> "When the Soviet Union fell, optimistic scholars believed the world had shifted inexorably in the direction of free markets and liberal democracy. Instead, the West gradually embraced bigger government and weaker social bonds, creating a fragmented society in which the only thing we all belong to, as President Barack Obama puts it, is the state." - Ben Shapiro

The steppes of Central Asia only get limited amounts of water. It's not nearly enough for farming, especially modern farming. So the Soviets dammed the rivers.

Sound like a good idea, you say?

Well, the Soviets thought so too. They diverted as much water as they could into their farming efforts.

But what happened downriver?

It used to be that the water flowed into the Aral Sea, which was one of the largest inland freshwater seas on this planet. But since the Soviet central planners diverted the water, the Aral Sea nearly completely dried up and disappeared[91].

Yes, that's right. An entire sea disappeared almost without a trace. Except for a few sad remnants of the once great Aral Sea, a desert is left in its place. To this day, you can still go into the area and see huge abandoned cargo ships sitting in the middle of a burning sea of sand.

A region covering several countries was affected. The local fishing industry was lost. An entire ecosystem has collapsed. There is no more evaporation from the long-gone sea, so the rainfall patterns of several countries have been permanently changed. This region was once healthy, vibrant, and ecologically diverse. But today, the area around the Aral Sea is becoming increasingly barren as the desert advances. Wildlife is disappearing. So is the farming and the grassland that once supported huge herds of cattle, horse, goats, sheep, and other livestock.

> "The Communist bloc of old was a study in the failure of failure. Losers in the Soviet economy were the people at the end of the long lines for consumer goods. Worse losers were the people who had spent hours getting to the head of the line, only to be told that the goods were unavailable." - P. J. O'Rourke

In short, the disappearance of the Aral Sea is both an ecological and an economic disaster. Although it's virtually unknown in the West, it is the largest ecological disaster to ever hit planet Earth since a meteor hit the Earth and wiped out the dinosaurs 65 million years ago.

And it is not only a man-made disaster, it was a completely centrally-planned man-mad disaster.

THE END JUSTIFIES THE MEANS

To the Marxist way of thinking, the end justifies the means. Anything that results in achieving the goals of central planners is ok. And that means *anything*.

If we look to history to show us how the power of central planners has been used in the past, we see numerous horrendous events like the massive famine in the Soviet Union that occurred during the winter of 1932-1933. This

---

[91] "The Disappearance of the Aral Sea", Vital Water Graphics, United Nations Environment Programme, http://www.unep.org/dewa/vitalwater/article115.html and "FactBox-Key facts about the disappearing Aral Sea", Reuters, June 23, 2008, http://www.reuters.com/article/2008/06/24/idUSL23248577.

famine was completely created by Joseph Stalin, the dictator of the USSR. It was completely avoidable. During the famine, about 8-12 million[92] people died.

How could such a thing happen?

In 1927, Stalin launched what he called a "revolution from above" to force the Soviet economy to erase all traces of the free market[93]. His goal was to collectivize all production into state-controlled and state-owned industries.

> "Death solves all problems - no man, no problem."
> - Joseph Stalin

The ownership of all private property was eliminated. If you were a blacksmith, your tools were no longer your tools. If you were a farmer, your land was no longer your land. If you ran a factory, well that wasn't yours either. It was all put in the hands of central planners. The central planners often lived thousands of miles away. They had achieved their role as central planners not because they were good at central planning, but because they were influential in the communist party.

The result was devastating. Farmers' livestock was taken, reallocated, and reassigned wherever the central planners thought they were needed. Unfortunately, the central planners in Moscow were not all that great at farming in places like Siberia or the Ukraine. Their misallocation of resources caused food shortages on an unprecedented scale. And even those who did have food were forbidden to eat their own crops. They had to turn everything over to the state for reallocation to central population centers.

> "Ideas are more powerful than guns. We would not let our enemies have guns, why should we let them have ideas?" - Joseph Stalin

Peasants reacted to collectivization by refusing to work, killing their own livestock, selling what they had and moving to the cities where they could get jobs and food, or by just dying.

For those who think such events might be a fluke, consider the Chinese example of Mao Zedong's Cultural Revolution. When central planning failed, Mao accused the bourgeois (businessmen and merchants) of sabotaging communism and introducing counter-revolutionary capitalism into the country.

---

[92] It's difficult to get an exact figure because the Soviets did their best to deny the famine ever happened and to suppress all records of it. Some estimates are as high as 28 million people dead.

[93] "Revelations from the Russian Archives: Collectivization and Industrialization," The Library of Congress, http://www.loc.gov/exhibits/archives/coll.html.

To "renew the spirit of the Chinese Revolution"[94], Mao launched a long series of violent attacks on anyone considered to be part of the bourgeois. The targets of the Cultural Revolution were often businessmen, merchants, intellectuals, or even just people who appeared to be intellectuals because they wore glasses.

Anyone thought to be spies, "running dogs", revisionists of the revolution, or hostile to Mao was beaten, imprisoned, raped, tortured, sent to "re-education camps", had their property seized, denied medical treatment, or just executed. Millions were forcibly displaced[95].

> "Reagan won the Cold War by first restoring America's economy and military and then staring down an economically weakened Soviet Union. He knew defeating Russia couldn't be accomplished without laying the groundwork." - Kathleen Troia McFarland

This doesn't even touch the atrocities of many other Marxists such as Adolf Hitler, Daniel Ortega, Nicolae Ceausescu, and so many others.

In all, Marxism is estimated to have killed between 85 and 125 million people in the 20th Century[96].

And all of this was considered to be for the greater good of mankind.

GOVERNMENT CAN'T FORESEE, PLAN, OR LEGISLATE INNOVATION

Now that we've seen what central control can do, let's take a look at what it can't do.

Central planners can never plan innovation. They don't know when it's coming. They can't legislate it and make it happen.

Many government officials have done their absolute dead level best to force innovation to occur. Not one has succeeded.

For example, Barak Obama wanted to move America away from fossil fuels and into green energy. To that end, he promoted an agenda of helping companies that were making solar energy cells.

[94] Kenneth G. Lieberthal, "Cultural Revolution," Encyclopedia Britannica, June 4, 2014, http://www.britannica.com/EBchecked/topic/146249/Cultural-Revolution.
[95] Harding, Harry. [1987] (1987). China's Second Revolution: Reform after Mao. Brookings Institution Press. ISBN 0-8157-3462-X
[96] Stéphane Courtois, "Introduction: the crimes of communism" in The Black Book of Communism, 1999, pp. 1–32.

One such company was Solyndra, which produced a new type of solar panel. Obama personally visited Solyndra and made sure they got a government-guaranteed loans to fund their business. This was in spite of repeated warnings that Solyndra might fail[97].

And it did. When market forces drove the unprofitable Solyndra out of business, the American taxpayer was on the hook for losses of over $800 million[98].

But of course it doesn't end there. In all, Obama's clean energy initiatives resulted in losses of somewhere around $2 billion of taxpayer money[99].

> "It is true that liberty is precious; so precious that it must be carefully rationed." - Vladimir Lenin

As much as Mr. Obama wanted to *make* green energy happen through government regulations and funding, he can't. No one can. Innovation can't be planned or legislated, no matter how hard anyone tries. No one can point to a single innovation that was ever planned by the government. Not one. Isn't it about time we stopped beating that dead horse?

PERSONAL PROPERTY

Marxism is based on the idea that property is theft. All property must belong to everyone. If you have something that is uniquely yours, then you have stolen it from everyone else. Therefore, the all powerful state must take control of your property. It must redistribute the wealth of the nation from those who have earned it to those who have not. It's the "you didn't build that" mentality of modern government flunkies.

Under Marxism, one of the main purposes of the totalitarian government is to train all human beings to let go of the idea of personal property. In fact, it was Karl Marx who famously said, "The theory of communism may be summed up in one sentence: Abolish all private property."

> "If history could teach us anything, it would be that private property is inextricably linked with civilization." - Ludwig von Mises

---

[97] Joe Stephens and Carol D. Leonnig, "Solyndra: Politics invused Obama energy programs", The Washington Post, December 25, 2011, http://www.washingtonpost.com/solyndra-politics-infused-obama-energy-programs/2011/12/14/gIQA4HlIHP_story.html.
[98] Sean Higgins, "Report: Taxpayer loss due to Solyndra may be as high as $849 million", The Washington Examiner, October 22, 2012, http://www.washingtonexaminer.com/report-taxpayer-loss-due-to-solyndra-may-be-as-high-as-849-million/article/2511398.
[99] The exact figure is difficult to estimate because these energy initiatives were funded through a number of programs.

Figure 5.1 shows why this can never possibly work.

Figure 5.1 This cat understand economics better than
any Marxist.

Figure 5.1 shows a picture of my daughter's cat. His name is Simba[100]. Notice that Simba is playing with something. Although it's difficult to see in the picture, his little cat toy is a small stuffed penguin. Simba loves this penguin. He carries it from place to place in his mouth. When Simba goes to bed, the penguin *must* be in the bed too or Simba will be upset.

The point here is that the penguin is the private, personal property of the cat. It's not my penguin. Nor is it my daughter's penguin (at least, not any more). It's *Simba's* penguin. And the critical thing here is that *Simba knows the penguin is his*. If you try to take it away from him, he'll either get sad and pout (yes, cats can pout) or he'll get mad and scratch you.

Why is that a big deal?

It's a big deal because an 11 pound ball of fur completely understands the concept of private property, which is the most fundamental idea behind any economic system.

If you are religious, you have to believe that God gave Simba the capacity to understand this idea and that this capacity to understand and want private property somehow helps Simba. It's part of God's plan that Simba should need to understand this or have a brain that's designed to understand this.

---

[100] I must sadly report that Simba is now deceased, thanks to local coyotes.

If, on the other hand, you believe in evolution, you have to believe that there is some evolutionary advantage to Simba having the ability to understand and want private property. Evolutionary biologists tell us that organisms quickly get rid of functionality that doesn't benefit them and their species. There's an example right inside every human being. It's called an appendix. I have no idea what an appendix used to be for. But whatever it did, it doesn't do that now. Humans don't need that task done any more. So appendixes are vestigial organs. That is, they don't do anything. They haven't yet completely disappeared, but they don't actually function. Our bodies eliminated that functionality because it was no longer necessary.

The ability of Simba to understand that that penguin is *his* and *no one else's* is something that Simba needs. Otherwise it would be long gone.

> "Private property began the instant somebody had a mind of his own." - E. E. Cummings

There are millions of years of evolution between Simba and human beings—maybe hundreds of millions of years. But the ability to understand the idea of personal property has persisted through all those eons of evolution. Even after all that time, it's still integrated into the design of both cat brains and human brains. Therefore personal property *has* to be an evolutionary advantage to organisms that understand that concept. If it wasn't, we would have lost it in the time that we evolved from cats to human beings.

Or, for the religiously-minded, Simba's brain shares many common design features with the human brain. If you believe that God intentionally designed both humans and cats, then you have to believe that there is some reason that He put the concept of personal property into that design.

Either way you look at this, there's a reason that both cats and humans can comprehend the idea, "This stuffed toy is mine and it belongs to no one else." And our reaction when things that we think are ours get taken away from us is the same as Simba's. We either pout or we get mad.

The problem with the Marxist idea of the complete communal ownership of all resources, property, and means of production is that it fights directly against literally eons of human evolution— or against the purposeful design of an all-knowing God. Either way, it's on the losing end of reality.

> "The dream of socialists, the Maximum Programme, has always been to eliminate the private property, the family and the nation state. With the private property they have not succeeded, but they continue on the path of destruction of the family and the nation." - Vladimir Bukovsky

Living in a world where everyone always smiles and is always happy to share and does not act out of self interest is not reality. You cannot force human beings into a world like that no matter how hard you try. It is to their advantage, speaking in terms of survival, to understand what personal property is and to want it.

The desire for personal property cannot possibly be eliminated from human beings. It's so integrated into our brains and our very beings that it's essentially a natural force, like the wind. And like the wind, it can be destructive. Anyone who's ever been in a tornado or hurricane understands exactly how destructive the wind can be.

But the wind is necessary for human survival. It brings the rain we need to live. And from ancient times, intelligent people have harnessed the wind to transport them around the world. They did it by innovation. They built boats with big sheets of cloth sticking up in the air. They called the sheets of cloth sails. They used their sailboats to go wherever they wanted. In fact, with nothing more than big pieces of cloth, they circumnavigated this entire planet.

The point here is that we can fight the forces of nature and destroy ourselves, as Marxists do, or we can harness the forces of nature and accomplish miracles. And there are many miracles we can accomplish if we are willing to innovate. In fact, I assert that we *must* innovate to survive and that the blockchain is the perfect platform for that innovation.

WEALTH CONFISCATION AND HURTING THE POOR

To get a clearer picture of why Marxism can't work, we need to understand what happens when the government confiscates wealth and personal property from those who have earned it.

DOWNTON ABBY

Downton Abby is a popular British TV series that shows the lives of the members of the aristocratic Crawley family in the early 1900s. The Elder Lord Crawley, the patriarch of the clan, is married to an American heiress. He had a title and land but little money. She had money but no title or land. So the match was good for them in terms of their time.

At one point in the series, the heir and co-owner of Downton Abby, Matthew Crawley, dies. Unlike a leech, which disconnects from its host when the host dies, the government actually keeps sucking on you after your death. Specifically, they tax the wealth that you left behind for your family.

Governments impose this ghoulish grab at you because they think it is unjust for you to pass the wealth that you have earned through your industry, hard work, and thrift on to your family so that your family might have a better future. They view it as unjust that your descendants might accumulate enough wealth to finance the purchase of their own homes (without getting loans from government controlled Fannie Mae and Freddie Mac), pay for their own educations (without getting government-backed loans), and invest in their own companies to create jobs for themselves and others. For some reason, it seems horrible to governments that families who are stable, thrifty, and hard working should reap the benefits of their own labor.

> "The only difference between death and taxes is that death doesn't get worse every time Congress meets." - Will Rogers

The idea behind the inheritance tax is that the government knows how to spend your money better than you do. This is in spite of the steady stream of news reports that we all see detailing massive government waste.

Politicians try to convince us that it is unfair to the poor that a rich person's descendants should get his or her wealth when they did not earn it. This is deception by misdirection. Even if the government takes the money, it never gets to the poor. In never helps them a bit. It goes into government coffers where politicians can use it to advance their own goals.

> "Congress can raise taxes because it can persuade a sizable fraction of the populace that somebody else will pay." - Milton Friedman

In any case, the Crawleys are thrown into an emergency because they must find a way to pay an oppressive inheritance tax without selling off the farmland around the estates. Selling the farmland would not only destroy the ability of the Crawleys to have an income, it would destroy the livelihoods of the many farmers that live on their estate (the farmers can't afford to by the land themselves).

The confiscation and redistribution of wealth is a foundational idea of Marxism. Socialists, communists, and other "useful idiots"[101] who believe that the only way you can acquire wealth is if you oppress and steal from others. Therefore, they openly support the maxim that "all property is theft." If you are successful, they believe you are a thief.

---

[101] These words are not mine. This is the term that Karl Marx uses for the ignorant masses that help communists attain their goals.

However, confiscating wealth has a human cost that Marxists never want to talk about. The Downton Abby series illustrates the inhumane nature of confiscating wealth quite clearly. The Crawley family not only is in danger of losing their own livelihood, so are all the farmers who live on the land. In addition, all of the merchants and other businesspeople whose incomes depend on the revenue they get from both the Crawleys and the farmers are also in danger. The entire local economy is on the edge of destruction.

The point here is that confiscating wealth has a very human and very devastating cost.

You might say that if the government takes the money, it will spend the money and that money will create jobs elsewhere. So all will be well in the end. The Crawleys and everyone else who depends on the estate can just go get new jobs.

Wrong.

First, that ignores the devastation that is wrecked on the lives of the farmers, merchants, and businesspeople on the estate because their functional local economy is arbitrarily destroyed by the government. It basically says their lives mean nothing.

> "To lay with one hand the power of government on the property of the citizen, and with the other to bestow it on favored individuals .... is none the less robbery because it is .... called taxation."
> - US Supreme Court in Loan Association v. Topeka (1874)

Second, government can never spend money as efficiently as private individuals and industries. Consequently, even if the government spends the money, fewer jobs are created than if it was spent by individuals and industries. There is a net decrease in jobs when government grabs money from us and spends it. For this reason, the government has never created one single job.

It's true that there are a lot of people who work for the government. But because of the net decrease in jobs that results from government confiscation of wealth and government employment, the net number of jobs that the government has created is actually negative. Government employees are a drag on the economy.

The upshot of all of this is that if the government forces the Crawleys to sell their farmland to pay the inheritance taxes, it not only deprives many people of their livelihoods, it makes it harder for those thus deprived to find a replacement source of income. Because there are fewer jobs in the economy due to government spending, the displaced farmers, merchants, and other businesspeople stand a very good chance of not finding a replacement job. Inheritance taxes actually penalize the poor.

And let's face it, the rich find ways to avoid the tax. When they do, they create a wasteful and economically draining industry of lawyers, bankers, and tax accountants that grow up in order to help them avoid taxes. Because money is being misdirected into this unnecessary expense, the economy is less efficient and creates fewer jobs due to less effective investment.

Even if the Crawleys do end up paying the inheritance tax[102] and they lose their livelihood, it's less likely to hurt them than the poor. They have the resources to start over in something new. They may lose their huge home, but they'll most likely just invest in something new and find another way to earn money.

The poor, on the other hand, who are displaced and whose livelihoods are ruined by this tax are unlikely to recover. They may fall into poverty from which their descendants never rise.

PARIS HILTON

Most of us have, unfortunately, seen or heard of Paris Hilton in the media. She is the granddaughter of a man who, through his hard work, careful planning, thoughtful spending, and thriftiness, built a massive fortune through such businesses as the Hilton Hotels.

Marxists LOVE people like Paris Hilton. She spends obscene amounts of money foolishly. She is famous for her sleazy, idiotic behavior. Marxists shout from the rooftops that wealth should not remain in the hands of a train wreck like Paris Hilton. Government should take it away. It should go to the poor.

> "Confiscation of private wealth does not make the public or even its agent, the government, rich. It does not create equality of wealth, but an equality of poverty." - Anonymous.

Wrong.

First, although the amount of money wasted by Paris Hilton is truly obscene, it's small compared to the amount of money that the Hilton family holds. Her wasteful spending is far less, and therefore far less wasteful, than say the

---

[102] Spoiler Alert! The Crawleys find a way to keep going without selling the farmland.

Environmental Protection Agency (EPA)[103] or the Department of Health and Human Services (DHHS)[104]. These two government agencies are actually depriving the poor of much more money that Paris Hilton could ever possibly squander.

Second, the vast majority of the Hilton money ends up in the hands of you and I already.

Say what?

It's true. The Hiltons don't keep their money as cash in a super-secret family vault deep under the estate. They put it in banks. The banks then lend that money to you and I for our cars, educations, houses, and so forth. Without the Hilton cash, the economy would be poorer than it is. There would be fewer jobs.

> "I do not believe that the Government should seek social legislation in the guise of taxation. If we are to adopt socialism, it should be presented to the people of this country as socialism, and not under the guise of a law to collect revenue." - President Calvin Coolidge

If the Marxists had their ways, the Hiltons would be poor beet farmers living by the shores of Lake Erie. But then thousands of people would not have money for the college tuitions, houses, cars, and businesses that they need. Hilton money doesn't stay with the Hiltons. It's circulated, invested, and re-circulated and reinvested. It enriches all of us. Taking that money and giving it to government central planners would result in far fewer jobs and lower standards of living for all of us. Is that what we want?

Wealth confiscation *always* backfires. It doesn't help the poor. It just helps whoever is in control of the government because it provides more money for them. The confiscation of personal property and wealth results in money being spent less efficiently and having less of a positive impact to the economy. We actually have more money for innovation, education, and so forth if we leave it in the hands of those who earned it–or even their profligate descendants.

---

[103] Rick Moran, "The EPA wants to watch you in the shower", American Thinker, March 18, 2015, http://www.americanthinker.com/blog/2015/03/the_epa_wants_to_watch_you_in_the_shower.html.
[104] DHHS wasted about $125 billion according to the Government Accountability Office. See Peter Fricke, "2014 Welfare Waste Dwarfs Forced Military Cuts By 25%", The Daily Caller, March 17, 2015, http://dailycaller.com/2015/03/17/2014-welfare-waste-dwarfs-forced-military-cuts-by-25/.

WILL POLITICIANS PROTECT US?

It is childishly naïve to believe that power-hungry, money-grubbing, self-serving, elitist politicians with Napoleon complexes will protect us from power-hungry, money-grubbing, self-serving, elitist businessmen with Napoleon complexes. They are birds of a feather. Of course they will flock together against us. And that's exactly what they are doing.

Calling for the government to do more to protect us against businessmen is like calling for more foxes to guard the henhouse. Political parties are complete sellouts. The different parties sell out to different special interests. That's the only real difference between them. They're not going to protect us at all. It's true that they do pass laws. But those laws are seldom good for us.

For example, the Affordable Care Act (ACA, or Obamacare) was billed as a way for the poor to get health care.

Poppycock.

The ACA simply puts the government in greater control of health care. The direct results was that *fewer* people were covered were covered than before the ACA was enacted. Literally millions of people lost their health care plans[105] in spite of Mr. Obama's now infamous promise, "If you like your plan you can keep it. Period."

It's true that now more people are covered than ever. That's a fact that Mr. Obama like to trumpet a lot. Unfortunately, virtually all of the people who lost their plans are covered by insurance that is more expensive and gives them fewer benefits–myself included. Mr. Obama like's to say, "We can't go back." But I'd LOVE to go back to the plan I lost as a result of the ACA.

It's also true that there are many people now who are getting subsidized health care. They effuse on social media how wonderful it all is now that they have affordable health care. But when you confront them with the fact that those of us who lost our previous plans are getting less *and* subsidizing their health care, they flat don't believe it. They can't imagine that anyone other than the ultra-rich are paying for their health care. They are completely disconnected from the real costs of what they're receiving.

---

[105] Robert F. Graboyes, "If You Like Your Plan, You Still Can't Keep It", US News and World Report, Spet 22, 2014, http://www.usnews.com/opinion/economic-intelligence/2014/09/22/under-obamacare-americans-will-continue-to-lose-coverage and Betsy McCaughey, "Another 25 million ObamaCare victims", The New York Post, January 14, 2014, http://nypost.com/2014/01/14/another-25-million-obamacare-victims/.

The only thing that the ACA really does is enrich insurance companies because it forces everyone to buy insurance. The government has failed completely to protect people from business. Quite the opposite. It has enabled the insurance industry to feed at the government trough on a much larger scale than ever before. And it has done so at the expense of middle Americans.

Marxists effuse endlessly about how fair and wonderful everything will be when the government protects us from mean, scary businesspeople. But where are they going to find these wonderful guardian angels to put into the government? Where are they going to find these kind souls who never put their own interests first and who are always kind and benevolent to you and I?

The reality is that politicians are just like the big, bad businesspeople that the Marxists want to protect us from. Politicians' first and foremost interest is themselves. They are interested in attaining wealth and power without actually having to work for it.

The Marxist ideal of a kind, loving, benevolent government that keeps people safe from evil businessmen is simply a fantasy that can't possibly exist in the real world. Fantasy Land always sounds like a great place to live. But actually moving there is highly problematic.

MYTH: "IT JUST HASN'T BEEN IMPLEMENTED RIGHT"

There are an increasing number of useful idiots who tell us that, "Marxism sounds good in theory, but it just hasn't been implemented right."

No. Just no.

There is only one way to implement Marxism, and we've already seen it done many times.

> "Strong as it looks at the outset, State-agency perpetually disappoints everyone. Puny as are its first stages, private efforts daily achieve results that astound the world." - Herbert Spencer

Why is there only one way to implement it? Because by definition Marxism requires a redistribution of all wealth. People who have wealth don't want to give it up. That means there has to be someone to take it away and someone to redistribute it. And that means there has to be central controllers with guns. And it also means that there has to be central planners to say who gets the "redistributed" goodies.

As we've already seen, central planners can't do what the free market does. They can't plan or create innovation. Therefore, the economy can't move improve. We become stuck in yesterday's economy. Without the freedom to innovate, there is an upper limit to the number of jobs that yesterday's economy can create. New products, services, and industries cannot be invented. Our standard of living doesn't grow. The economy can't move forward and it stagnates. We've seen this in every single Marxist country that was ever created.

In addition, the redistribution of wealth of course leads to resistance. The only possible outcome of resistance is an all-powerful totalitarian state that can force the will of central planners on everyone.

> "A lie told often enough becomes the truth." - Vladimir Lenin

For example, under Marxism, if there are more college professors than the central planners think the economy needs, the central planners must reassign some of them to other work. So some might become plumbers or factory workers or whatever else the central planners think is needed.

This is the only possible outcome of following the Marxist philosophy. There is no other way to implement it. We've seen the one and only way Marxism can ever be implemented and it's resulted in the largest bloodbaths in history. Do we really need that history to replay itself in our streets in order to understand that Marxism can't possibly work?

CRONYISM AND MARXISM ARE IDENTICAL

We saw in Chapter 4 that the free market, in its purest form, died a long time ago. What we have now is a mix of the free market, cronyism, and Marxism.

Unfortunately, cronyism and Marxism are the same thing. Don't believe me? Let's do a side-by-side comparison and see.

The Power and Wealth Structures of Cronyism and Marxism

Figure 2.3, back in Chapter 2, showed how our economy works. I'll repeat it here in Figure 5.2 for your convenience.

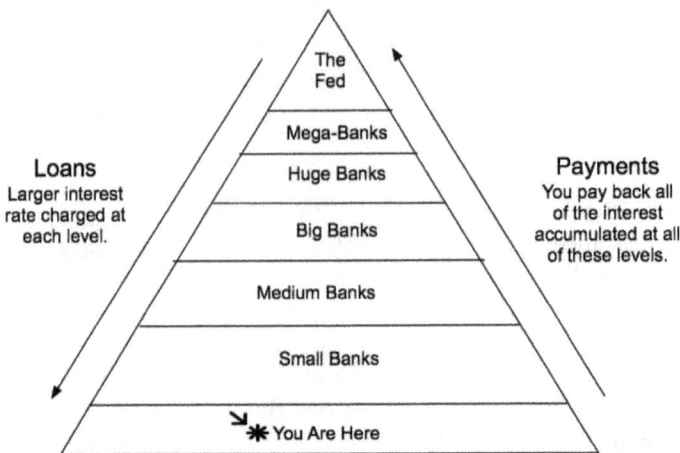

Figure 5.2 Cronyism concentrates wealth and power from the many to the few.

Recall that because the bankers lend money in a pyramidal structure with the ultra-wealthy elites at the top and the rest of us further down, they essentially control all the resources and production of our society. It all has to be used to pay them back.

The unelected bankers at the top of this pyramid decide how much unemployment there should be, the interest rate on your mortgage, the price of housing, the value of the dollars in your bank account, and virtually all other aspects of the economy.

As was previously noted, the Fed's process of manipulating the economy concentrates both wealth and power to the bankers at the top. As former Congressman Ron Paul states, "It is increasingly obvious that the Fed's post-2008 policies of bailouts, money printing, and bond buying benefited the big banks and the politically-connected investment firms. QE is such a blatant example of crony capitalism that it makes Solyndra look like a shining example of a pure free market!"[106]

Figure 5.3 shows how Marxism works. Does it remind you of anything?

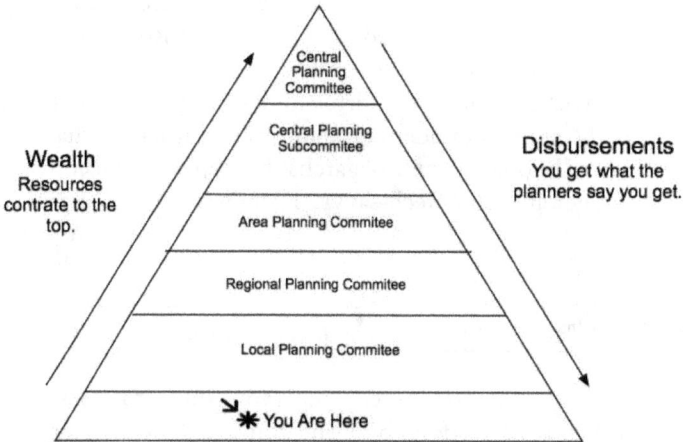

Wealth
Resources
contrate to the
top.

Central
Planning
Committee

Central Planning
Subcommitee

Area Planning Commitee

Regional Planning Commitee

Local Planning Commitee

☀ You Are Here

Disbursements
You get what the
planners say you get.

Figure 5.3 Marxism gives a little but takes much more.

Unlike cronyism, disbursements which hands out money first and then takes more back, Marxism starts by taking everything of value away from everyone. It concentrates the resources directly to the central planners, who plan every aspect of your life. The central planners decide how much you should make, what kind of apartment you should live in, what kind of clothes you should wear, what you should do for a living, and so forth. In that respect, the difference between cronyism and Marxism is only one of degree. Either way, money, power, and resources concentrate into the hands of the people at the top. Those at the bottom are left to fight each other for whatever crumbs they can get ahold of.

The effects of Marxism are the same as the effects of cronyism. Either way, control over all wealth concentrates to the few at the top. They centrally plan the economy and apportion out what they think you and I should have. They then disburse that wealth to us in such meager portions that we end up completely dependent on them—which is exactly what they want.

---

[106] Ron Paul, "Ron Paul: Federal Reserve Steals From the Poor and Gives to the Rich", RonPaul.com, November 18, 2013, http://www.ronpaul.com/2013-11-18/ron-paul-federal-reserve-steals-from-the-poor-and-gives-to-the-rich/.

The only real difference between Marxism and cronyism is the speed with which the effects take place. Marxism immediately forces the centralization of power and control over the economy onto everyone. Cronyism does it slowly, but in the long term the effect is exactly the same.

If you doubt the equivalence of Marxism and cronyism, take a look at what's happening in Russia and the United States today. In Russia, even though the communists are largely gone, the country is mostly ruled by a group of rich oligarchs. The average citizen has virtually no input into how the country is run. This was the situation under communism and it hasn't changed with the fall of the Soviet Union. The new oligarchs that replaced the communist party bosses are younger and more tech-savvy. There is a somewhat greater amount of economic freedom under the oligarchs. But otherwise, things have not changed that much as far as the political and wealth structures are concerned.

> "I predict future happiness for Americans if they can prevent the government from wasting the labors of the people under the pretense of taking care of them." - Thomas Jefferson

In the United States, we see exactly the same thing to a lesser degree. A recently published study[107] done by Princeton and Northwestern universities concluded that America is no longer the democratic republic that was purposefully designed by the Founders. It has in fact become an oligarchy run by wealthy elites who are routinely able to enact legislation that directly conflicts with the will of the public and is counter to their interests. The average voter in the US has virtually no input into national policy. And in fact, I know from personal experience that voters who contact their representatives in Congress are routinely laughed at when they express their dismay at the representatives votes.

If cronyism is allowed to continue in the long term, it's cumulative effect will be no different than Marxism. It's just a slower process, but the process is exactly the same in the long term.

The Same Two Problems

Cronyism and Marxism are so similar precisely because they suffer from the same two problems. They both use the same banking system and they both are economic monopolies. The fact that Marxists spout anti-capitalist rhetoric and hatred is meaningless. In every socialist or communist country that has

---

[107] Martin Gilens and Benjamin I. Page, "Testing Theories of American Politics: Elites, Interest Groups, and Average Citizens", Perspective on Politics, Cambridge Journals, September 18, 2014, http://journals.cambridge.org/action/displayAbstract;jsessionid=54C3FA20371F7AB5A4E3F53B1E0AB086.jo urnals?fromPage=online&aid=9354310.

ever existed, the government has always issued fiat money from a government central bank—just the same as every capitalist country. The fact that they put Lenin's or Mao's picture on the money means nothing. The money itself and the banking system behind it are identical to the government fiat money and the central banking systems used worldwide.

Whether the country is communist, socialist, or supposedly capitalist makes no difference. All countries that issue government fiat money through a central banking system will have the exact same two problems in their economies; their money will be broken and so will their banking system.

A country can call itself communist, socialist, or capitalist. But the long-term results of its economic policies will be exactly the same. All power and wealth will concentrate to the elites at the top. In the so-called capitalist countries, the process is slower because there is nothing that will drive the concentration of power and wealth as fast as Marxism. But fast or slow, the destination is the same.

The only thing that can change this relentless concentration of wealth and power from the many to the few is rethinking money itself. And that is what we'll talk about next.

PART 2

REINVENTING MONEY

6 THE SOLUTION: APPLY FREE MARKET PRINCIPLES TO MONEY ITSELF

The way we solve many of the problems plaguing our civilization is to end the government's monopoly over currency and create a market of freely competing currencies. Essentially, we must apply the principles of the free market to money itself.

COMPETITION IN CURRENCY

Calling for denationalized, private currencies that compete in the free market is a pretty radical idea. And unfortunately, I cannot claim it as my own. Competing free market currencies were first proposed (as far as I know) by the Nobel Prize winning economist Friedrich August Von Hayek.

Hayek first proposed the idea in the mid-1970s and wrote multiple books on the subject. I strongly recommend that you read his books <u>Denationalization of Money: The Argument Refined</u> and <u>A Free-Market Monetary System and The Pretense of Knowledge</u>. These two books were probably the most radical volumes on economics that were ever published.

Creating a system of competing currencies will do what competition does with all free market products. It will produce a better product. Because we're talking about money here, let's be specific. A free market system of competing currencies will create forms of money that are more stable, hold their values better, are easier to use, and have better features that anything we have today.

People may very rightly ask, "Don't we already have that? Lots of governments print lots of different types of money."

No. All governments print the same type of money. It doesn't matter whether it bears a picture of a president, queen, prime minister or a rhinoceros (yes, there *is* money with a rhinoceros on it), it's all the same thing. All of it is government fiat currency issued through debt by a central bank. The decorations on the front and back don't matter. It all has the same features.

"The great trouble is that money wasn't allowed to develop. After two or three hundred years of the use of coins, governments stopped any further developments. We were not allowed to experiment on it, so money hasn't been improved, it has rather become worse in the course of time." - F.A. Hayek

All governments inflate their money to the extent that the public will let them. So all money everywhere suffers from inflation. All money everywhere is issued by a central bank. So it all concentrates wealth and power from the many to the few.

As Hayek stated, "I am more convinced than ever that if we are ever again going to have a decent money, it will not come from the government."[108]

So that means that we have to create our own money if we are going to have stable, valuable monetary systems. But the good news is that we can. And the money we create is better than anything the government can offer.

WHY USE PRIVATE CURRENCIES?

Why would anyone want to use a currency issued by a business, nonprofit organization, or an individual rather than a currency issued by a country?

There are basically three reasons.

1. Free market currencies provide choice. They give businesses, organizations, and individuals the ability to use the currency they prefer.

2. Free market currencies hold their value. Their buying power stays constant over time. Government-issued currencies always undergo inflation. They are never stable.

3. Free market currencies are the basis of economic democracy. Political democracy cannot survive in the long term where there is a governmental monopoly over currency.

If a free market of competing currencies will create meaningful change, then we should look at how such a market can be created.

---

[108] F.A. Hayek, A Free-Market Monetary System and The Pretense of Knowledge, Ludwig von Mises Institute, Auburn, Alabama, p 16.

CREATING A PRIVATE CURRENCY

In most developed countries it is illegal to print your own currency. In many of them, this was once not the case. Private currencies–bank notes–were often issued by banks in Western nations for at least a century. In most cases, it was longer. Not one of the nations where bank notes were in circulation could show any detrimental effects from their availability[109].

The way to create a currency is to use the blockchain to follow the pattern of bitcoin and its derivative currencies, the so-called altcoins (alternative coins). Under this system, an individual or group of people writes a distributed software system that no one runs. It functions according to known rules and algorithms. Instead of an organization issuing the currency, the software issues a digital currency on its own. This is perfectly legal in almost all countries.

Money created this way is purely digital. There is no physical manifestation of it. There are no coins and bills, just bits. But that's ok. Most people really don't care about coins and bills any more if bits can be used to achieve the same thing.

That is not to say that we should just use bitcoin. I'll talk more about bitcoin later and explain why it can't be the basis of a fully functional economy. What I'm saying here is that having the blockchain means that we now live in an age where anyone can make a software system, put it out for general use, and have the software system issue a currency. The currency then operates autonomously according to a set of known rules.

Autonomous currency systems are a radical idea. There are worries about them. People wonder if they can be hacked. Or they may be afraid that they can be exploited in some way that is detrimental to everyone else. Those are valid concerns. So are any other concerns you may have about digital currencies. I'll address those shortly. For now, the important point is that we have come to a point where we can build custom currency systems. We really can have a system of competing currencies that produce stable, easy-to-use money. There is nothing that restricts us from applying free market principles to money itself.

---

109 It is true that there were "wildcat banks" issuing bad currencies in the US in the 1800's. However, this was the result of too little competition, not too much. Big banks were prohibited by law from opening branches in small towns. So the banks in small towns could abuse the system without the fear that their customers would leave for more stable currencies.

ANYONE CAN USE A PRIVATE CURRENCY

But how can regular people use private currencies?

Any citizen of almost any country can already use multiple currencies to do business, buy, or sell. In many countries, merchants can already accept foreign currencies when they buy and sell.

Note

In a few countries, it is illegal to buy, sell, or fulfill contracts in any currency but the national one. However, these countries are the exception rather than the rule.

Under the model pioneered by bitcoin, no one individual or organization issues the money. Instead, people create a software system that issues money and processes transactions. The entire system is distributed, meaning no one owns it, no one is in charge, and no one can take control. Anyone can participate if they decide that it's in their best interest to do so. The currency system is a distributed, decentralized, self-organizing system (DDSOS) just like the Internet.

A DDSOS currency system can be designed to be used in tandem with other currencies. Merchants may accept multiple currencies as easily as they accept one.

Multicurrency buying and selling already occurs along the borders of many countries. For example, along the US-Canada border, merchants on both sides accept both currencies. Merchants typically charge a conversion fee for dealing in a foreign currency. That is because they expect to ultimately spend or hold their own local currency. However, if the companies they buy from, their supply chain, also accepts multiple currencies then there is no need for them to charge the conversion fee.

WHAT GIVES VALUE TO CURRENCY?

At this point, it's sensible to stop and ask, "What gives money its value? Why should anyone believe that private currencies have any value at all?"

The answer is that there are four things make money valuable.

1. The supply of money matches the demand for it.

2. The money is trusted.

3. The features of the money appeal to large numbers of people.

4. Some entity (the government) uses force to *make* people accept the money.

Let's examine each of these ideas in turn.

MATCH SUPPLY TO DEMAND

In centuries past, currency was all in precious metal. It was the value of the metal itself that gave value to the currency—or so we once thought.

There are multiple examples of times in history when the value of the metal and the value of the currency became disconnected in a way that provided a positive result. We will present two.

In 1879, Austria was using silver coins for its money. The value of silver took a sudden downturn. This caused the value of Austria's silver coins to decline precipitously. Carl Menger, an Austrian economist of the time, advised the government as follows.

> Well, if you want to escape the effect of the depreciation of silver on your currency, stop the free coinage of silver, stop increasing the quantity of silver coin, and you will find that the silver coin will begin to rise above the value of their content in silver. (A Free-Market Monetary System, F. A. Hayek, p 17-18. Ludwig von Mises Institute.)

Menger's answer was simple; issue fewer coins. He basically told them that they should only supply as many coins as there is a demand for. If the supply goes down, the value will go up.

The government followed Menger's advice and decreased the number of their silver coins in circulation. The result was exactly as he predicted. The value of an Austrian silver coin quickly returned to its face value, rather than the value of the silver in the coin. It saved the economy. But it also proved that currency does not derive its value from precious metals.

> "Money is made possible only by the men who produce." – Ayn Rand

Exactly the same thing happened in India fourteen years later. They took the same course and got the same result.

What does this tell us?

First, currency doesn't need to be backed by a precious metal to make it valuable. It is simply when you limit the supply of a currency to match the demand that it retains its value. If the amount of currency does not exceed the demand for that currency, then the currency will have value and people will accept it.

People may ask the very valid question, "What about the gold standard?" For a long time, many countries were on a national gold standard. The US was one such country. That is, there had to be one dollar's worth of gold held by the government for each paper dollar that it issued. In theory, one could go to Fort Knox (where all the gold was kept), knock on the door, and demand to exchange paper dollars for gold. In reality, no one ever did that. Eventually, that was also made illegal in order to consolidate government power over money.

> "Money is worth what it will help you to produce or buy and no more." – Henry Ford

During the Nixon administration, the US government realized that there was really no connection between the gold it held and the currency it issued. Therefore, it abolished the national gold standard and began printing money in whatever quantities it wanted[110].

Some people today call for a return to the national gold standard. This is not necessary. All the gold standard did was to limit the government's ability to rampantly print money. But that genie is out of the bottle, so to speak. It cannot be put back in. The US will never return to a gold standard because politicians love inflation far too much.

> "One penny may seem to you a very insignificant thing, but it is the small seed from which fortunes spring." – Orison Swett Marden

Fortunately, we don't need a gold standard. And anyway, there is not enough gold in the world for every country to be on a gold standard. But it doesn't matter. In a free currency market, competition forces currency issuers to do their best to keep their currencies at a stable value. They need only match the supply of their currencies to the demand to accomplish that.

---

[110] It's true that the departure from the gold standard was caused by a geopolitical and economic situation that was ongoing at the time. But the government never would have dumped the gold standard unless the disconnect between the value of gold and the value of the dollar was clear.

TRUST

Another important aspect of currency that gives it value is trust. If the issuer of a currency has the trust of a large number of people, its currency will be used. As long as the issuer proves that it can maintain the buying power of the currency at or near a fixed level, that trust will continue to grow. As it does, the currency will see wider and wider adoption.

If a currency issuer is ready to be completely straightforward with its customers about when it issues money and how much it issues, then that currency issuer is more likely to succeed in the free market. Transparency engenders trust. Transparency enables people to objectively evaluate the currency and decide for themselves whether they think the currency will maintain its value over time.

In a free currency market where many currencies are available for use, anyone can select the currencies from the issuers they trust. It doesn't matter whether that issuer is their bank, their church, their civic group, or their labor union.

We can see what happens when people *lose* trust in their currency over and over in history. Currencies that go into hyperinflation don't always get there because the government prints an unending river of money. Government only has to print enough money for people to lose their trust in it. After that, everyone tries to exchange it for something else so the value of the money plummets. Let's make it clear that the government must overprint *a lot* of money before people will get to that point. However, after that point is reached, it doesn't matter whether the government prints a little currency or a lot. The trust is gone and the money is valueless.

> "Trust becomes solidified when words consistently back up by deeds." – George David Miller

FEATURES

Like any other product, people will use a certain type of money if it is easy and convenient. If people find that a particular kind of money is hard to use, they won't use it. They'll prefer something easier.

FORCE

Governments use force to make people use their money. The free market does not. In fact, it can't. It can only compete on the basis of supply, trust, and features. I assert that when you compare these three characteristics of free market money to any other form of money, you will find free market money to be the best kind of money currently available.

COMPETITION EQUALS EFFICIENCY, USABILITY, STABILITY, AND SAFETY

Long experience with free markets shows that when there is healthy competition among the producers of a product or service, the price of producing that product or service goes down. In addition, the efficiency of producing it goes up.

In the case of free market money, competition will ensure that only the most stable forms of money survive. If there are stable currencies around, no one will want the currencies that fluctuate wildly or steadily lose value over time. People want to keep the value of their savings and not lose it to inflation. So government currencies, and any other currencies that lose value, will disappear. Either that, or their issuers will have to adopt sane monetary policies that maintain the value of the money they issue.

Based on free market experience, it is reasonable to predict that the competition in a free currency market would produce a currency that is highly stable, efficient to use, and retains its value over a long period of time. Stability, efficiency, and value over time are the primary features that most people desire in a currency.

A free currency market also provides greater safety for consumers. If a currency begins to lose its value, people will quite naturally move into other currencies to protect their savings. Likewise, businesses will protect their investments by diversifying their currency holdings. No longer will anyone be tied to the fate of any single currency. They can protect themselves from the effects of economic instability by exercising their free choice in the currencies they use and hold.

BUSINESSES NEED STABLE CURRENCIES

To continue to do business over time, companies must have a stable currency. For instance, when a bank makes a loan during an inflationary period, the value of each dollar that they receive when the loan is paid off is less than the value of the dollars that they loaned. It is true that they receive more dollars than they lent. However, because the individual dollars have been devaluated, they make significantly less profit than they would if the money had a stable value.

If the money that the bank lent out was stable and kept its value, banks could make more loans because they make more profit. In an inflationary economy, banks make fewer loans than they otherwise would. Because banks can make fewer loans, there is less opportunity for new businesses to start and for existing businesses to expand. The economy produces fewer jobs and unemployment may even result. Which, of course, causes government to intervene again by generating more inflation.

In wage contract negotiations, business is the big loser when inflation occurs. While it is true that they may give less value in wages over time than they received in labor, the workers' unions are not dumb. They know that they must build in a cost of living increase into their contracts. They typically take their members out on strike when the company does not want to give such a large wage increase–often resulting in lost profits for the company. The effect of strikes can ripple through a large portion of the economy.

If management does accede to the demanded cost of living increase, the company loses again if inflation is not as large as anticipated. In that case, they are paying too much for their wages. The result is higher costs, less business, and fewer new jobs.

Also, inflation and higher wages are costs that businesses must pass along to customers. Higher costs usually drive demand for their products and services down. That is, they will do less business because of higher costs. They may become less competitive than companies in other countries that have a more stable currency or are not as unionized.

In every respect, inflation is bad for business. Business must have a stable currency.

CITIZENS NEED STABLE CURRENCIES

The average citizen also needs a currency that does not inflate. People do not want the buying power of their money to decrease over time. No one likes the fact that when US president Barak Obama took office, the price of gasoline over most of the nation was approximately $1.80 and by the end of his first administration, it more than doubled. The effect on the economy was profound. The price of virtually everything increased as the rise in gas prices impacted all aspects of business.

It isn't only gasoline that has been hit by such out-of-control inflation. In less than one lifetime (mine), the price of one cup of milk purchased as part of a school lunch has gone from 4 cents to 40 cents. The average consumer can cite similar examples each time they go shopping. They know quite clearly that the money they earn is being eaten away by inflation very dramatically. It will only get worse.

The value of a worker's life savings goes down over the lifetime of the worker. Saving money is rapidly becoming an exercise in futility. Because the buying power of money is going down, workers have little hope of a decent retirement.

In every respect, inflation is bad for consumers. They must have a stable currency.

WON'T GOVERNMENT BAN PRIVATE CURRENCIES?

Given the tendency of governmental power structures to perpetuate and protect themselves, it's logical to ask whether government will ban private-issue currencies. The answer of course is that some of them will.

In most democracies, it is perfectly legal to hold and use a foreign currency. The same is true of currencies issued by decentralized monetary systems.

Note

In a few countries, it is not legal to buy, sell, or fulfill contracts with foreign currencies. The Qbit is a non-national currency, so it is legally treated as a foreign currency in many nations or as a commodity (like gold) in others.

But some repressive governments may outlaw the specific use of non-national currencies because they fear the competition to their own money. I encourage all citizens of all countries to behave in a legal and lawful manner.

Experience shows, however, that simply banning private currencies won't work. First, there are many forms of private currencies. For example, if you shop in a store that offers a rewards card, you are using a private electronic currency that is specific to that store.

So the quick answer is that banning private currencies would have a much larger effect than intended. It is even possible that banning private currencies could make discount coupons commonly used for shopping illegal.

Also, history shows that when people have a chance to use good money, they will use it even if there are legal penalties. During the French Revolution, the paper money issued by the revolutionary government quickly became worthless. Nevertheless, it was against the law to buy and sell using gold coins or the currencies of other countries. The common Frenchman faced a simple choice. Either use worthless French paper money and starve or use other forms of money and maybe get arrested.

As time went on the revolutionary government attached the death penalty to the use of any currency besides their own. In spite of that, the common people continued to use other forms of money. If they did not, death was certain. If they used other forms of money, they might get the guillotine or they might not. At least they had a chance at living.[111]

If people are faced with government currencies that will eventually wipe out their life's savings, they will gladly choose to use a non-national currency instead if that currency will keep its value over time. At least they will have a chance at something other than starvation. It is likely that people will even risk jail to preserve the benefits of their work for themselves and their children.

Warning

These statements should not be taken as an encouragement by the author or the publisher to disobey the laws of any nation. They are simply a prediction of human behavior based on the lessons of history. The author and the publisher strongly discourage the use of private currencies where it is not legal.

Although governments may initially look upon free currency markets with alarm, we hope that they will soon see that the long-term benefits to all

---

[111] For more information on this, please see "Fiat Money: Inflation in France", Foundation for Economic Education, Irvington-on-Hudson, New York, 1959, pp. 75-89.

people will ensure domestic prosperity and stability in ways that nothing else can.

LEGAL TENDER LAWS

To implement a monetary policy, government must give itself a monopoly over the nation's currency through the use of so-called legal tender laws.

WHAT IS LEGAL TENDER?

Legal tender laws basically say that you must accept the "coin of the realm" whenever it is offered. That means that you cannot refuse to be paid in US dollars in a transaction in the US. As a result, if there is a better currency than the one provided by the government, people will always pay with government money and hold the better currency. Sellers eventually don't want to sell in the better currency because anyone can be pay them in inferior legal tender at any time. In this way, the use of other currencies disappears. Bad government money pushes out other, more stable currencies.

Legal tender laws are written precisely with the intent of pushing all other currencies out of use within a given country. Every national government wants its citizens to use its own currency regardless of whether or not that is the best currency to use. In this respect, government seeks its own interests, not that of its citizens.

YOU CAN USE PRIVATE CURRENCIES

In most countries, legal tender laws do not prevent you from paying in other currencies. They also do not prevent you from accepting other currencies.

As I mentioned previously, along the US-Canadian border, stores on both sides accept either currency. They charge a conversion fee if you pay in the other country's currency. But that is because they buy their goods primarily within their own national borders using their own national currencies.

Suppose a store in the US accepted a private currency just the way it can accept Canadian dollars. Wouldn't it also charge a conversion fee? It wouldn't *have* to. If the store's supply chains also accepted the private currency the store would have no reason to charge a conversion fee. They could simply pay in private currency.

But there's a problem with accepting any currency but that of your own nation. For example, stores in the US *must* accept US dollars if they are offered. It does not matter whether the store prefers a private currency or Canadian dollars or Japanese yen. They *cannot* refuse US dollars. And of course, Canadian stores *must* accept Canadian dollars no matter which currency they prefer. So the long and short of it is, you cannot accept *only* private currencies even if you want to.

As a merchant, you can shift the balance away from legal tender by offering discounts for the currency you prefer. Stores along the US-Canadian border typically charge a 3%-5% conversion fee. Therefore, if you set your discount for paying in a private currency to 3%-5%, you will be in line with common industry practices.

If your profit margins do not allow for a discount for your favorite private currency, simply raise the US dollar price by 3%-5% and do not raise the price in the private currency. This accomplishes the same thing.

When writing contracts, legal tender laws mean that you can be paid in dollars even if you write the contract for payment in a private currency. That also can be balanced out. You simply add a clause that states that the payer must bear all the costs of the contract and that if they pay in dollars (or whatever your local legal tender is), they must pay an additional conversion fee so that you can convert your payment in dollars to the private currency you want.

Adding a conversion fee clause stays in effect even if the contract is breached. Suppose you make a loan in a private currency and specify that the debtor pay you back in that same currency. Imagine also that the debtor breaches the contract. You take him to court and the judge agrees that the contract was breached. She awards you payment, but of course it is in dollars (if you're in the USA). Because of the conversion fee clause, the debtor must also pay the conversion fee for you to convert your damages into the private currency, which is what you wanted originally.

So the long and short of this particular discussion is that you *can* use private currencies in most countries in spite of legal tender laws if you know how.

CREATING A MULTICURRENCY WORLD

Issuing private currencies that compete against government-issued debt-based currencies is a straightforward way to apply free market principles to money itself. But living in a multicurrency world has the potential to become ungainly and overwhelm consumers. So let's move on to the next chapter and see how such a situation would work.

7 DIGITAL CURRENCIES

Recent developments in blockchain-based digital money have demonstrated that currencies do not have to take the form of physical cash. It is possible to create and use money that is entirely digital.

WHAT IS DIGITAL CASH?

Digital cash is information that represents monetary value in just the same way that physical cash represents monetary value. Digital cash is encoded with advanced cryptographic codes that make it impossible to forge. It is actually easier to make counterfeit paper money than it is to make counterfeit digital cash.

The cryptographic encoding used to create digital cash also makes it impossible to spend the same digital cash more than once. The system will not validate both transactions.

DIGITAL CASH IS NOT A CREDIT CARD

Before going much further with the idea of digital cash, I want to emphasize that digital cash is not the same as a credit card. A credit card is a financial instrument that is a promise to pay, like an IOU. It is a loan from the credit card company to you. When you use a credit card, the credit card pays the seller you're doing business with using its own money. It then charges you interest and expects you to pay back that loan over time.

On the other hand, digital cash is just like physical cash in that it contains value in and of itself. Digital cash is not a promise to pay. When you pay a seller with digital cash, the seller is paid. It's just like paying with physical cash.

Also, credit cards do not give you any privacy. The credit card company sees what you buy, who you buy it from, and how much you pay. Digital cash, like physical cash, is completely private. The system doesn't know who you paid or what you

> "I think that the Internet is going to be one of the major forces for reducing the role of government. The one thing that's missing, but that will soon be developed, is a reliable e-cash..." – Milton Friedman, 1999

bought. It just sees a transfer of funds from one anonymous account to another. There is no way to trace the transaction to you at all.

When you have a dollar bill, you can't look at the dollar bill and get any information about who had it last, what they bought, and what the price of the item was. Digital cash is the same way. It keeps no information about you, so you have the same privacy as physical cash.

DIGITAL CASH IS NOT A DEBIT CARD

It's also important to state that digital cash is not a debit card. A debit card is a financial instrument that banks use to transfer money from one person to another. In that respect, it's no different than a check. It just moves the money faster than checks do.

Like credit cards, debit cards give you no privacy. The bank sees what you buy, who you buy it from, how much you paid, and when the transaction occurred.

Once again, digital cash is private, just like physical cash. It keeps no record of who had it previously and what they did with it. You don't know who used the dollar bills in your wallet before you received them. No one who uses them after you do will ever know that you used them. Digital cash is exactly the same way.

DIGITAL CASH IS CASH THAT'S DIGITAL

Let's stop a moment and draw an analogy. When we digitized printed documents, the paper disappeared. We don't use it any more. We still have the documents, but they're completely digital. If we digitize money, the same things happens-it makes the bills and coins go away.

One more analogy: we digitized music and that made the records and tapes go away. We still have the music. But now we need digital music players to listen to it. The same is true of digital money. When you digitize money, the coins and bills go away but you need a digital device to store your money on. For most people, that will be their computers, their tablets (like an iPad), or their phones. All you need is digital wallet software. So just like you store your documents and music on devices, you store your digital cash on some kind of computerized device.

Digital cash has almost all of the properties of physical cash. Specifically, it is fungible, recognizable, limited, private, liquid, costly to obtain, valued, useful for trade, and useful as a store of value.

When you use physical cash, possession is ownership. Digital cash has that same feature. If you lose your digital cash, it's exactly the same as losing physical cash. The digital cash system can't help you recover it any more than you can knock on the doors of your country's mint and ask them to replace cash that you lost.

Also, digital cash can be stolen just like physical cash. The digital cash system can't help you recover it.

Note, however, that there are ways of backing up your digital cash in case of hardware failures or theft. And we'll discuss those later in this book.

WHY USE DIGITAL CASH?

So why would anyone want to use digital cash rather than physical cash?

CONVENIENCE

There are many reasons to prefer digital cash over physical cash. One reason is convenience. Digital cash is as easy to use as any other form of money. Digital cash can be stored on electronic devices, kept on cards with magnetic strips, and even printed to paper.

In addition, you can have debit and credit cards issued by banks that enable you to spend digital cash in exactly the same way that you spend your local currency. This reduces privacy, just as it does when you use credit cards to pay in US dollars or other national currencies. But many people area already willing to sacrifice a bit of privacy for the convenience.

You can use a simple camera phone—even a very cheap one—to get prices from stores that accept digital cash. This is the case even if the price is displayed in your national currency. It's simply a matter of pointing your phone at a bar code on the item's price tag and using the correct app to display the price in the currency you're using. So using digital cash can be as convenient as using physical cash.

PRIVACY

Another reason to use digital cash is that it is private, as I already pointed out. A physical cash keeps no record of who has used it and what they bought. Digital cash is the same.

Privacy is especially important in the digital age. In current forms of online shopping, there is no privacy. Digital cash enables private transactions to take place even when you are shopping online.

Because sellers are paid immediately, just like they are when you use physical dollars, there is no need to identify yourself when you pay with digital cash. No store cares who you are if you pay with physical cash because they already have their money. No one makes you show an ID when you pay with cash.

Digital cash is the same. No one cares who you are because you are paying with cash, just the same as when you pay with physical cash. They have their money so they don't need anything more from you. Digital cash maintains your privacy.

STABILITY

The biggest factor in preferring digital cash over other forms of currency is that the digital cash does not inflate if the system is created properly. Digital cash systems can be designed to hold a reasonably constant value over time. So no matter what's happening in the economy, your savings will not disappear due to inflation. You can rest assured that your money will continue to have a reasonably constant value over your lifetime.

HOW DO I USE A DIGITAL CURRENCY?

To jump into the digital cash economy, use the following steps.

STEP 1: DOWNLOAD A DIGITAL WALLET

To enable yourself to buy and sell with digital cash, you must first download the digital wallet software from whoever issues the digital cash. Typically, the digital wallet software will run on your PC, Mac, or mobile device. So you can pay with cash directly from your computer or handheld device.

As digital cash systems become common, more and more companies will offer online wallets. These wallets are different than the ones you download and run on your own devices. They are hosted on a web site that is always available from anywhere you can get an Internet connection. Because they are web-based applications, the only things you need in order to access them is a device with a web browser, an Internet connection, your user name, and your password.

Typically, you would use an online wallet in more or less the same way you would use a digital wallet app on your computer or mobile device. That is, you can pay for goods and services online or in stores. The only real down side to an online wallet is that you must have an Internet connection to use it. You are also entrusting a third party (the company that runs the wallet service) to keep your cash safe for you. That may or may not be a good idea.

STEP 2: BUY OR ACCEPT DIGITAL CASH

Next, you can use your digital wallet to exchange some of your local money for digital cash. You can keep this digital cash in your wallet on your computer or download the mobile version of the wallet and put your digital cash on your phone.

Increasingly, digital cash providers allow you to use a distributed online currency exchange to buy your digital cash. This eliminates the middleman because you are buying directly from someone who already has the digital currency. This is about as safe as buying something on Craigslist. If you want to take a safer route, you should use an online currency exchange that you trust.

You can also sell a product or service and accept digital cash as payment. This is probably the easiest and safest way to get digital cash.

STEP 3: GO SHOPPING

When it's time for you to shop using your digital cash, there are two ways to do it. First, you can shop in a physical store. Second, you can shop online. Let's take a look at both of these shopping experiences and see how they would proceed with digital cash.

Shopping in a Store

At the store, start the wallet software and point the camera at the price tag, which has a 2D bar code similar to the one in Figure 7.1.

The bar code on the price tag probably contains the price in US dollars (if you're in the US). The reason for this is that most merchants will want to sell to both those who use digital cash and those who do not.

Figure 7.1 A 2D Bar Code

You point your device's camera at the 2D bar code and your digital wallet software shows you the price in the digital cash of your choice. If you hold several currencies in your wallet, the wallet software can show you the price in each one.

The store can also accept multiple digital currencies. In addition to the price of the item, the bar code on the item can hold information about which currencies the store accepts. Your wallet can look at the list of the currencies that the store accepts and match it to the currencies you've got in your wallet. So your wallet won't bother to show you the price of the item in currencies that the store won't accept. You see how much the item is in every currency that you have that the store accepts. If you don't have any of the digital currencies that the store wants, then you can just pay in your national currency, such as US dollars if you are in the US. That's because the wallet software can store your credit card or debit card information and complete the transaction using your credit/debit card instead of digital currencies. So in the end, you're almost guaranteed to be able to pay for your item as long as you have money.

Just like when you're shopping with US dollars, you then take the item to the cashier to pay. Payment is done with your wallet software. At the cash register, the cashier rings up your purchase just like always. The cash register is just a device that runs the right software. Sellers may use digital wallet software to perform transactions. However, they will find this inconvenient if they sell goods regularly. For that, digital cash issuers are increasingly providing cash register software that runs on most mobile devices. So the cash register can be an inexpensive tablet PC that supports an external screen. Many tablet PCs support external screens, so the store might even use a tablet.

Once the cashier totals up the purchase, the cash register generates a 2D bar code containing the invoice for the sale. It displays the bar code on the external screen, which faces the buyer. If you are the buyer, you point your cameraphone at the 2D bar code on the screen. Your wallet software automatically goes to the location specified in the 2D bar code and gets the invoice. It then displays, "Bob's Corner Grocery: 2 Woolongs" (if you're paying with a digital currency called Woolongs).

To complete the purchase, you simply select "Accept Purchase" and your wallet software pays the cashier. Confirmation of the transaction takes about as long as confirmation of a credit card purchase.

Please be aware that this scenario is the lowest common denominator. Most phones–even phones as inexpensive as $15–have Bluetooth, near field communication (NFC), or Wi-Fi built in. If your phone has any of these capabilities, the cashier can send the invoice to your phone directly. Then you don't have to worry about using the camera at all.

It is also possible to have debit and credit cards that enable you to pay in digital cash. In that case, shopping is just a matter of swiping your card.

Note

Customers should be aware that paying with a debit or credit card is not as private as paying with digital cash. However, it is safer because paying with digital cash is just like paying with physical cash. So you need to decide whether privacy or safety is more important to you.

Shopping Online

As digital cash becomes common, online stores can display prices in the currencies that the store accepts. So when you begin shopping online, web sites that accept digital cash should display item prices in that currency and any other currency you choose.

If you want to buy an item with your digital cash, you simply pay with your digital wallet. The web site will "talk" to your wallet and display an invoice. When you approve it, the merchant is paid.

DIGITAL CURRENCY FOR ECONOMIC DEMOCRACY

As I've stated before, having a free market of competing currencies creates a competitive situation in which only the most stable, valuable, and convenient forms of money survive. Digital cash systems can be designed to make it easy to use multiple currencies at once.

Applying free market principles to money itself creates a healthier, fairer, and more humane economy. It eliminates inflation and fractional reserve banking, which drive people into poverty and create artificial levels of competition for limited supplies of currency. Consumers can spread their risk out over multiple currencies to help ensure their financial futures.

Digital currencies provide an easy way to use multiple currencies with the same convenience that we now have when using one currency.

BITCOIN PAVES THE WAY

I've mentioned bitcoin a few times without really saying what it is. Because bitcoin technology is so fundamental to digital cash systems, it's important to dive into bitcoin and understand the basics of it before we go much further.

WHAT IS BITCOIN?

Bitcoin is a completely distributed currency and transaction system that was invented by a programmer who went by the pseudonym Satoshi Nakamura. Its great innovation is that it uses a method of storing transaction information which guarantees that the same digital cash can't be spent twice. This method of information storage is the blockchain.

BITCOIN IS BASED ON THE BLOCKCHAIN

In more technical terms, bitcoin's blockchain is a data structure that validates and maintains the state of a completely decentralized distributed database system. Using the blockchain, decentralized systems can operate with the assurance that system data will always be in a valid state as long as bad actors do not gain more that 51% of the system's computational resources.

162

You can think of the blockchain as a set of folders on your hard drive. The folders, which represent the data blocks, are all daisy-chained together one after the other. Only the most recent folder is open. The others in the chain are all locked. In fact, the farther you go back in the blockchain, the more securely the folders are locked.

Everyone on Earth could potentially have a copy of the blockchain. The magic of the blockchain is that as long as your computer is connected to the Internet, your copy of the blockchain will stay in sync with all the other copies that exist everywhere on Earth.

THE BLOCKCHAIN IS A TRANSACTION LEDGER

The blockchain stores a public ledger of all of the transactions that happen in the bitcoin system. Using the blockchain, the bitcoin system can validate any transaction that anyone wants to make. It basically proves that your bitcoins are valid and that they are in your account.

With the blockchain, the system can prevent double spending. That is, if I pay Alice three bitcoins, I cannot turn around and pay Bob with those same bitcoins. I don't own them any more. Even if I have a digital copy of them, I spent them. The system sees Alice as the owner now and not me.

The blockchain's ledger is completely public. But that doesn't violate your privacy. The ledger uses anonymous account numbers, similar to numbered Swiss bank accounts. All that it records is that X amount of bitcoins moved from account A to account B. It doesn't track what was bought, where it was bought, who paid for it, or who sold it. All of that is private.

THE BLOCKCHAIN IS MORE IMPORTANT THAN BITCOIN

Even though bitcoin's blockchain is a complex and highly technical topic, I'm taking time to provide you with a brief overview of it because it's so critical to digital cash systems. In fact, **the blockchain is more important than bitcoin itself.**

Bitcoin is the first application of the blockchain. But bitcoin and the blockchain are two different things. Bitcoin is a good effort at a currency. But it cannot be the basis of a fully functional economy. I know that there are those who would disagree with that. In fact, there are some who look at bitcoin with a religious fervor. But as good as bitcoin is, **it is not as important as the blockchain**. The blockchain has uses that are far beyond bitcoin.

Let's draw an analogy. I remember a time when the Internet was used mostly for email. Email was a great thing. We still use email. Email has been hugely impactful. But the Internet that moved the email around has proven itself to be far more important than email.

Well bitcoin is like email. It's great, but it's not what all the hoopla is about. Just as the Internet has proven itself useful for far more things than email, the blockchain will prove itself useful for far more things than bitcoin.

The blockchain enables anyone to create distributed, decentralized, self-organizing monetary systems. Remember that whole DDSOS stuff I was talking about in previous chapters? This is where it all becomes important. As you'll soon see, the blockchain enables us to completely redesign all of our monetary systems. And the impact of that cannot possibly be overstated. By redesigning our monetary systems, we are fundamentally redesigning our civilization. That is not an exaggeration, and I'll prove it before the end of this book.

HOW DOES BITCOIN WORK?

Bitcoin operates by validating every single transaction as it comes into the system. The transactions are collected together into a block and crypographically signed. This prevents anyone from forging anything in the blockchain.

Because anyone can have a copy of the bitcoin blockchain, anyone can process transactions for the system. When you install a copy of the bitcoin blockchain software, it automatically finds all of the other computers in the system that are processing transactions. It gets a copy of the blockchain and "magically" keeps your copy in sync with all the other copies.

A Decentralized Transaction Ledger

Although the blockchain may not sound like a big deal yet, it really, really is. Because all copies of the blockchain are automatically kept in sync, there doesn't have to be any central control of the system. So there's no one target for hackers to attack.

Even if a hacker does break into one copy of the blockchain and alter it, he gains nothing. All of the other copies of the blockchain will reject his changes. Anything he does will be seen as invalid and no other copy of the blockchain will believe his copy.

The Problem with Centralized Systems

Until the advent of the blockchain, any networked computer system that offered any kind of product or service had to have a central control point. Think about it. Suppose you use Skype for making calls to your friends across the Internet. Skype is a centralized system. Microsoft owns Skype. If Microsoft wants to, it could track all of your calls that you make with the system. If the government wanted to, it could demand that Microsoft turn over all of the records of your calls to them (this actually happened in the NSA domestic spying scandal).

In addition, Microsoft could sell all of your call data to a third party. To my knowledge, Microsoft has never done that. But it could. And there would be nothing you or I could do to stop them.

Centralized systems are not private. They give their owners control over information about you. They are targets for hackers. They are targets for government control.

The Problem with Decentralized Systems

Decentralized systems are more private than centralized systems. But they have their shortcomings too. The main problem with them until now has been keeping information in a consistent state across the entire system.

Before the blockchain came along, it was nearly impossible to build a distributed, decentralized monetary system. That was because all the computers in the system had to keep track of transaction information. But there was no real way to keep all of the copies of the transaction ledger in sync. Banks and other big companies had solutions to that. However, they all depended on some level of central control. And that more or less defeats the purpose of a DDSOS.

The Blockchain Solution

The blockchain eliminates the need for central control in distributed computer applications. It automatically keeps all copies of the system's transaction ledger in sync with every other copy. If there are errors, it fixes them automatically. If there are deliberate forgeries, it rejects them.

Anyone can join the system and provide computing services for transaction processing. So the system can grow as large as it needs to. And the people who provide services to the system get paid for it automatically by the system.

So the blockchain enables us to create DDSOSs that can do nearly anything and that can pay for themselves as they go. And I'll demonstrate why that's the best thing since sliced bread, to coin an old phrase, throughout the rest of this book.

And by the way, I know I keep telling you that I'll explain things later and that can be annoying. But there is so much here, and this technology is so pivotal, that I can't present it all at once. So please be patient with me and let me wave my hands occasionally and say, "More on this later." Because I promise I'll make it worth the wait.

THE LIMITATIONS OF BITCOIN

One of the primary complaints of those who don't like the bitcoin system is that there can never be more than 21 million bitcoins. This means that, as a currency, bitcoin is permanently deflationary. Each coin always grows in value. So prices in bitcoins are always changing.

Another problem with bitcoin is that it requires third party currency exchanges to enable people to buy into the system. Mt. Gox, one of the first and largest bitcoin exchanges, was poorly run and went into financial collapse. Many of its customers lost money when it did. An exchange in China operated well for a few months, and then disappeared with all of its customers' bitcoins. The total value of the theft was in the millions of dollars.

Finally, bitcoin, for all its greatness, is a first-generation (1G) system. There are things that it was simply not designed to do. It's important to understand that bitcoin is open source technology. Anyone can get the bitcoin source code and improve it. There is a very dedicated community of programmers doing just that. So bitcoin is a technology that is improving. Nevertheless, it is a first-generation system.

THE SECOND GENERATION IS COMING

So that's an extremely brief introduction to bitcoin and the pioneering blockchain technology. But wait, there's more. In fact, the second generation (2G) of blockchain-based currency systems is being born as of this writing. These 2G systems can leverage the blockchain to provide custom economic systems that can truly solve massive problems that have plagued mankind for centuries. We are really at the beginning of something so radically different that it is hard to overstate the magnitude of what's coming.

So let's move on to Chapter 8 and see what all the fuss is about.

8 THE BLOCKCHAIN

In previous chapters, we saw the problems that our world faces and how those problems developed. Along the way, I asserted that applying free market principles to money itself will help us solve those problems. I also put forward the idea of issuing private currencies using a new technology called the blockchain and provided a brief introduction as to what the blockchain is.

In this chapter, I'll explain how the blockchain works. But don't worry. Even though the blockchain is a highly technical system, anyone can understand the basics of what it does and its operation. There are only a few essential concepts that the average person has to understand in order to see what the blockchain means in their lives.

Before this chapter is done, I'll also briefly discuss some of the potential impacts of the blockchain on our economy and on our liberties, since that is the focus of the book. As we go forward, I'll continue that theme with examples of how the blockchain can literally revolutionize our civilization.

But first things must be first. Let's dive in to how the blockchain works.

HOW THE BLOCKCHAIN WORKS

The details of the blockchain's operation are very, very technical. In order to understand them, you need a strong background in Computer Science and advanced mathematics, as well as a basic knowledge of encryption.

Fortunately, the blockchain is very much like your car. Or your iPhone. Or most of the things you use each day. You don't really have to know much about how your car works in order to use it. You've got to know that you have to put gas in it, change the oil periodically, and make sure the tires are inflated properly. Almost everything else can be handled by your mechanic.

The blockchain is the same way. If you know a few basic things about what it does and how it works, you can use it for all kinds of things. And those basic things are actually pretty simple.

1. The blockchain is a distributed ledger.

2.   It processes transactions.

3.   It runs on computers provided by miners.

A DISTRIBUTED LEDGER

As I touched on in Chapter 7, the blockchain is very much like the ledger an accountant keeps for a business. A business's ledger keeps track of all of the money that comes in and all of the money that it pays out.

The blockchain works essentially the same way. When you use the blockchain, you have a wallet that creates "accounts" that are very much like bank accounts. The accounts in your wallet hold coins. The coins are your money.

The blockchain's ledger keeps a record of all of the coins you receive and all of the coins you pay out.

Because the blockchain is distributed, there are many copies of it all over the world. This is good. It means that no one can hack the blockchain because if one copy of the blockchain becomes different than all the others, the others just reject the copy that's been tampered with. And no one can hack all of the copies of the blockchain because they can't get access to all of them.

The blockchain is called "the blockchain" because it stores blocks of transactions. Only the newest block of transactions needs to be verified. After another block is added on, the old blocks become more secure. Why? Because in order to hack an old block, you've got to hack every one of the newer blocks in the blockchain first. The longer the blockchain gets, the more secure its ledger becomes.

So the blockchain is a chain of linked blocks of transactions.

You may wonder, "If lots of people have a copy of the blockchain, can't everyone see how much money I have?"

The answer is yes and no. It's yes because people, in fact, can see how much money is in your account. It's no because no one knows the account belongs to you. So it's like having a numbered Swiss account. People can see money going in and out of it. But no one knows who's spending the money.

Now let me stop and qualify what I'm saying here. The pseudo-private accounts I've discussed so far is exactly how bitcoin works. But second generation (2G) blockchain systems are moving toward so-called "zero-knowledge proofs"[112] that make the blockchain even more private than a numbered Swiss bank account.

TRANSACTIONS

The purpose of the blockchain is to perform transactions. Whenever a transaction is performed, the blockchain verifies it. Specifically, it verifies that the same money is not spent twice.

If I pay some coins from my account to you, then the system won't let me spend those same coins again down at my corner grocery store. In other words, it prevents *double-spending*. Double-spending is a form of fraud. And one of the primary reasons the blockchain exists is to prevent it.

As the blockchain system verifies each transaction, it inserts the transaction into the newest block in the blockchain. It then performs an *confirmation* on the entire block.

For bitcoin, the confirmation can take as long as two hours (that is rare). Most of the time, it takes 2-10 minutes to get a confirmation on the bitcoin blockchain. This is one of the main reasons that bitcoin can never become the basis of a worldwide economy[113].

More recent blockchain systems are overcoming this difficulty. For instance, the Qbit system, which I am in the process of building, has a confirmation time of about 30 seconds. You may rightly assert that even a confirmation time of 30 seconds is too long in today's world of high-speed transactions. You're right. But fortunately the Qbit system, and other 2G blockchain systems like it, is designed to accommodate the integration of higher-speed transaction systems into them. In other words, you can put your coins into "escrow" in a 2G blockchain, issue digital "cashier's checks" against them, and then spend the cashier's checks as rapidly as you like. Later, in a slower-speed transaction, you can redeem all of the cashier's checks you have by presenting them to the blockchain and getting coins for them. This is actually

---

[112] An example is Zerocash. http://zerocash-project.org/how_zerocash_works.html
[113] There are also other reasons why I assert that bitcoin cannot be anything more than an experiment. Others would hotly disagree. This is a major source of contention, so-called "flame wars", in the blockchain community.

very similar to how a bank's high-speed transaction systems already operate. So this is a very natural service for banks to provide.

MINERS AND MINING

But wait. Who actually runs the blockchain?

Anyone can download the blockchain software and contribute their computer to the network of computers that process the blockchain. People do that with the intent of making money.

You have to be fairly technical to be able to run the blockchain on a computer. First of all, the computer on your desktop is not powerful enough. You have to build or buy a special computer that costs anywhere from $2,000 to $6,000, depending on how much computing power you want to add to the network.

Why would anyone spend that kind of money to join the blockchain? Because the blockchain system is designed to pay for itself. The people who put a computer on the network are called *miners*. Every miner gets a reward for solving a cryptographic puzzle. The solution to the puzzle is what secures the blockchain.

Remember just a bit ago when we were talking about the fact that each block in the blockchain must be confirmed? The miners do that when they solve the puzzle. Whoever solves the puzzle first gets some coins. So every miner is motivated to solve the puzzle first and confirm the validity of the transactions. In other words, every miner in the system is motivated by money to make sure the system is always telling the truth. The only way to corrupt the system is if you control more than 51% of the system's computing resources. But that's impossible.

For example, if Google used *all* of its massive computing power to try to corrupt the bitcoin blockchain, it would only have about 5% of what it needed in order to do that. Yes, all of the combined computing power of Google amounts to just 5% of the total computing power in the bitcoin blockchain.

In short, because each miner has a copy of the blockchain and all miners are financially motivated to ensure the system tells the truth, it's virtually impossible to make the blockchain lie.

In the bitcoin system, and most first-generation (1G) blockchain systems, the miners get all of the new money. Whenever they solve the puzzle and perform the confirmation on a block of transactions, they get paid. That's how money gets into the system. The miners then spend the money or exchange it for other currencies, such as US dollars, with those who want to buy coins.

2G systems may also have other ways of getting payouts. But either way, the blockchain always automatically pays the miners because the miners provide resources to the system.

If there are too many miners, any one miner won't make enough money to cover the cost of electricity. So some of the miners drop out.

If there are not enough miners, those who are already mining will make big money. That attracts other miners.

In other words, blockchain systems can automatically grow to the size they need to be. They don't need any outside entities to control their growth. They use free market forces to gain the resources they need without any outside interference.

THE BLOCKCHAIN'S IMPACT

It's hard to overstate the potential impact of blockchain technology. So let's go through a few examples to get an idea of why the blockchain is so critical to our future.

REMEMBER THE OLD MEDIA?

To understand just how big a deal the blockchain is, let's use an analogy. Up until the early 1990s, our worldwide media was totally centralized. Less than 100 people worldwide were the ultimate arbiters of what we saw and heard in our media.

Today, with the advent of the Internet in general and the World Wide Web in particular, that has completely changed. Everyone can become a content producer. Everyone can access the media and gain a worldwide audience.

If you doubt that, just fire up your browser and search YouTube for "Potter Puppet Pals". A group of high school students made their own puppets based on the Harry Potter characters. They put up a puppet stage in the basement of one of their houses and started making silly, creative videos of themselves doing Harry Potter-oriented puppet shows. What happened next was a New Media sensation. Potter Puppet Pals became one of the first amateur-produced videos to go viral across the world. Everyone knows the ridiculously hilarious songs from the Potter Puppet Pals. Anyone can watch them and nearly every kid does at some point or other.

Please don't miss the magnitude of the example of the Potter Puppet Pals. In an era dominated by multimillion dollar TV shows and movies produced by a multibillion dollar industry that was controlled by some of the richest and most powerful people on earth, a group of high school kids in a basement somewhere in Middle America was able to generate as much or more attention as the most popular movies of the time. Those kids were able to get more viewers than many of the most popular TV shows of the day. That was exactly the kind of threat that gave media company executives nightmares. But at the same time, it was a tremendous opportunity for millions of creative people all across the world. And it was all possible because new kinds of infrastructure (the Internet and the World Wide Web) became accessible to everyone.

The blockchain has this same potential. It does for money, banking, finances, and law what the Internet, the World Wide Web, and YouTube did for entertainment.

"OF THE PEOPLE, BY THE PEOPLE, AND FOR THE PEOPLE"

Because the blockchain allows anyone can build distributed systems with the blockchain that don't require central organizations to control them, it's possible to build large, complex systems that perform tasks we commonly assign to the government or that are commonly done only by large corporations.

Why is that important?

They Pay You

Right now, there exists mountains of data created by you and I. It goes into the hands of Big Business executives who find ways to monetize it. Google is the prime example of this, but there are other companies following Google's lead.

Google knows everything you search for. They keep a record of it. They use it to build a profile of you that they sell to advertisers.

Got a Gmail account? Google also reads all of your emails. It knows who your friends are, who you work for, what medical conditions you have, what your political affiliation is, and what church you go to (or whether or not your atheist). Google tracks what news you look at. It probably also knows what books you read even if you don't buy the books from Google.

But with the blockchain, we can build distributed, decentralized systems that enables us to keep our data private. So if anyone in Big Business wants it, they have to pay for it. Yes, they must pay for it. And that's exactly the kind of thing that keeps them up at night.

Open Internet

Let's take another example. The blockchain enables us to create an Internet that is completely peer-to-peer. No Internet service providers (ISPs) needed. The Internet can be reborn into something that no one controls and no one owns–a completely self-organizing system. It really is possible. An Internet like that can be the basis of peer-to-peer systems that are impossible right now.

The blockchain enables us to decentralize our world so that you and I have more control. It also enables us to compete more directly with large corporations.

Open Marketplace

As another example, what if you could build a system that did exactly what Amazon does that is designed like the Internet? That is, anyone could use the system to sell stuff but there was no corporation running it all.

How would such a thing work? In such a system, anyone could use the blockchain to post something for sale. People would use the blockchain to search for and order products. They would also use it to pay for the things that they buy with digital currencies.

Anyone who wanted to could hook their own computer to the system by attaching it across the Internet. They would do that because the system would pay them for letting it use their resources. If there were too many resources, people wouldn't be paid enough and they would no longer contribute their computers to the system. If there weren't enough resources in the system, it would automatically pay more for the computers it had access to. In an effort to make some money, more people would let the system use their computers. In other words, the free market would match the size of the system to the demand for it. No big corporation needed.

Future Opportunities

But it gets better. Suppose we look ten years into the future. Lots of companies are developing self-driving cars right now. Imagine that in ten years self-driving cars will become safe enough so that they can make deliveries on their own. I don't know if that will really happen or not, but let's imagine that we're in 2026 and self-driving delivery cars and trucks are common.

Now imagine that you decide to write a blockchain-based system, called DeliveryCoin, that enables anyone to offer the services of their delivery vehicles. Miners look at your system and say, "This is great! I'll mine coins on that blockchain." So they do.

Now you advertise your blockchain a little and people say, "Wow, this is great! With this system, I can get my deliveries made cheaper." So they use your blockchain to place orders for delivery trucks. They buy DeliveryCoins from the miners and others who hold them. They use those coins to pay for their deliveries.

I see your blockchain and say, "Wow, this is great! I own a self-driving, refrigerated delivery truck. I'll connect it to that blockchain so that it can get orders and make deliveries." So I do. My self-driving, refrigerated delivery truck gets its first order, pulls out of my driveway, and off it goes to haul its first load.

When the delivery is complete, I get paid in DeliveryCoins. I use the blockchain to sell the DeliveryCoins for dollars. We all live happily ever after.

Oh but wait, there's more to this story. What *really* happens is that when people pay, they pay my self-driving, refrigerated delivery truck, not me. My self-driving, refrigerated delivery truck then sends me whatever portion of the money I've told it to and keeps the rest of the income in a secure place in the cloud. I take my portion and sell it for dollars. But the truck keeps the rest. Why?

Imagine that my self-driving, refrigerated delivery truck continues to take orders through the DeliveryCoin blockchain. It realizes it needs gas, so it goes to the gas station, gets filled up, and pays for the gas using its own money.

Suppose my self-driving, refrigerated delivery truck gets a flat on its next delivery, it calls for a repairman. Mr. Repairman comes out and fixes the flat. The truck pays him. The truck also gets its own oil changes and other maintenance done on its own and pays with its own money.

My self-driving, refrigerated delivery truck is really getting successful. In fact, the demand is so brisk that my self-driving, refrigerated delivery truck can't keep up. So it buys another self-driving, refrigerated delivery truck with money it has saved up (I've authorized it to do that) and deploys the other truck out on the deliveries it can't cover. It's actually possible that my truck could eventually buy me a whole fleet of trucks, each of which are managed by the blockchain.

THE BLOCKCHAIN IS IN 1987

Wait, wait. Let's put on the brakes here. I've given you some pretty heady examples that sound almost like fantasy. The reality is that all of them are possible, but maybe some of them are not possible *right now*. We're only talking about the *potential* of the blockchain right now.

Make no mistake, we're in the very early days of the blockchain. Remember how the Internet was in 1987? No? Well I do. I've been in the high tech industry since 1981.

In 1987, the Internet was a quiet neighborhood. It was very small. There was email. But it was slow. It might take an hour for your mail to reach the other side of the world. But to us, that was fantastic.

And there was FTP, which let you transfer files from one place to another. But you had to do it right or you could really cause a problem. True story: a university professor of mine came to class one day and told us that he'd taken down the entire Internet all across the US Eastern Seaboard by trying to ship a file to a colleague that was too large.

How large was too large?

Two megabytes.

Yes. Two megabytes and the file broke the Internet.

Well, figuratively speaking, the blockchain is in 1987 right now. We're just starting out. Just like there was barely a functional Internet in 1987, there's barely a functional blockchain operating right now.

But remember, by 1994, the Internet was an Official Big Deal in the computer industry. And by 1997, it was everywhere. That's just 10 years.

The blockchain will go forward in the same way. There will be a gold rush of companies trying to use the blockchain for everything. Some will hit it big. Some will go bust. There may even be a "blockchain bubble" in which people overhype and overinvest in the blockchain like they did during the DotCom bubble in the late 1990s. However, that won't stop the blockchain any more than it stopped the Internet. By 2026, you and everyone else will know exactly what the blockchain is. And it will be an Official Big Deal in your life.

But right now, in 2016, the blockchain isn't much more than bitcoin. Some people are rapidly building expansive systems on top of the bitcoin blockchain. But bitcoin's blockchain wasn't built for that. So there are individuals and groups—and I am one such individual—who are rethinking the design of the blockchain. We are trying to build blockchains and blockchain systems that can grow, adapt to change, and become the basis for the complete infrastructure of a real economy[114].

THE BLOCKCHAIN, BANKERS, AND LAWYERS

With digital currencies you and I could exchange money without a bank. We don't need banks for home loans or insurance companies for insurance. We can build systems that enable us to provide each other with loans and insurance. We don't need Wall Street for venture capital. We can scale the process of getting venture capital down to the smallest entrepreneur. Lending and investing can become something that everyone can do themselves. Or they can hire a money manager to do it for them. Either way, the blockchain enables us to scale the power of Wall Street down to the average Joe on Main Street.

---

[114] For more information, please see www.qbitfederation.org.

Some think that blockchain technology will eliminate bankers and lawyers, and we'll see why over the rest of the book. The reality is that bankers and lawyers will change what they do. There will be mountains of new opportunities for them. They'll still have valuable functions in the world created by the blockchain. But what we know as a bank today will only slightly resemble what a bank will be in a blockchain-driven world.

Joi (Joichi) Ito, director of the MIT Media Lab, expressed his opinion this way. "Here's my prediction," Ito said, "I think that blockchain will be to banking and law what the Internet was to media. I think that we'll survive it, but we're going to be fundamentally different."

OTHER BLOCKCHAIN USES

It's important to remember that the blockchain is not just a global, decentralized ledger. In a more general sense, it's a completely decentralized database that can store any information and keep all copies of itself in sync.

So what? What's the big deal?

The big deal is that there are nearly limitless uses for the blockchain that have nothing to do with money, finances, privacy, or liberty. This book is focused on the economic aspects of the blockchain and how we can use it to create massive new monetary opportunities for ourselves, as well as how we can build economic infrastructures that reinforce our privacy and liberty. But the applications for the blockchain go far beyond that. I just won't be covering them in this book.

In short, there is no real limit to the killer apps that you can build on the blockchain. For example, the blockchain is the perfect tool for an automated copyright or trademark system that doesn't need to be run by the government. Yes it's true that the courts would still need to get involved sometimes. But on the whole, copyrights and trademarks can be handled on the blockchain.

On the blockchain, you can record marriages, land ownership, or even what house bought power from what power source. You can also have a shared, immutable history for mankind that can't be altered for political correctness or convenience. You can store the entire curriculum for a free education. It's even possible to store computer programs that anyone can download, buy, or use for free.

DO WE REALLY NEED THE BLOCKCHAIN?

You may sensibly ask, "Do we really need a new way of doing that stuff?"

The answer is a resounding, "YES!"

As we saw in earlier chapters, government is out of control. Increasingly grabby agencies are appropriating more and more of our privacy and liberties. The blockchain enables us to eliminate many of those government agencies completely. They simply become obsolete.

Also, because of the current economic woes in the world today, paths out of poverty are disappearing. The blockchain is a tool that we can use to completely reverse that. We have a chance to provide ourselves with a prosperity revolution unlike anything in the history of the world. As huge and pie-in-the-sky as that claim sounds, it is absolutely true.

Our technical society is moving toward what people in the computer industry call the Internet of Things. It's a level of interconnectedness that is beyond anything we have today. Our banking system simply wasn't built for that. In a world where nearly everything is connected to the Internet and exchanging money in very small transactions, it's literally impossible for banks to handle the resulting load. When you're talking about trillions, yes trillions of real-time transactions taking place all the time. There is no way that our banking system can scale to that level of business. It's just impossible.

On the other hand, a world of interoperating blockchains can handle the Internet of Things just fine.

If we're going to move forward, protect our liberties, and provide real paths out of poverty, then we *must* have the blockchain or something very similar.

BLOCKCHAIN-BASED CURRENCIES

In Chapter 6 I presented the idea of private currencies. In Chapter 7 I covered digital currencies. It's time to examine how a blockchain-based currency can function and how they solve problems that were previously insurmountable.

First and foremost, blockchain-based currencies free us from having to depend on central authorities like the Federal Reserve Bank for our money. It lets us escape from economic monopoly.

How does they do that?

The answer depends on which blockchain currency you're talking about. 1G blockchains use algorithmic currency production. 2G blockchains use much more sophisticated methods.

ALGORITHMIC CURRENCY PRODUCTION

Most blockchain-based currencies that exist today have a limit to the number of coins that the system will mint. With bitcoin, the limit is 21 million. There will never be more than 21 million bitcoins. Other coins have higher limits. They all produce money at a steady, predictable rate according to a known algorithm.

The nice thing about algorithmic currency production is that it happens according to a set of rules. For instance, because there is no Federal Reserve Bitcoin Bank, no one can suddenly decide to produce lots more bitcoins and inflate the value of the coins. If you like the rules by which the coins are produced, then the particular blockchain currency you're looking at can be a good fit for you. If you don't, you can just shop around until you find a currency you think is better. That's the nature of the free market; you have a choice. You are free to start using a currency you think will hold its value and free to stop using one that you think won't. The free market is freedom.

Having a choice in currency is also the basis for economic democracy. You vote with your wallet for the currency you think is best. No one has the right to steal that choice from you and force you into an economic monopoly that concentrates power and wealth to themselves.

AUTONOMOUS CURRENCY PRODUCTION

For 2G blockchain currencies, there is a much greater freedom in how the currencies are produced. Because I'm most familiar with it, I'll use my project, the Qbit, as an example to explain what I mean.

Unlike bitcoin, the Qbit system has no upper limit on the amount of money it can produce. The Qbit system is able to scan its own blockchain and determine the demand for currency at any given moment. It is artificially intelligent and it can learn. Because the blockchain keeps a perfect record of the variations in the price of the Qbit, the demand for it, the volume of transactions that are occurring, and the velocity of the transactions, it learns to manage the money supply far better than any human or group of humans is able to do. As time goes on, the Qbit system learns to manage the money supply in ways that stabilize its value. In other words, there is no real inflation or deflation[115].

That's right. Unlike any other form of money, the Qbit can monitor its own economy and tell how much currency should be in circulation. As a result, it can expand or contract the money supply automatically.

If the Qbit system sees that inflation or deflation is occurring, it will make adjustments in the money supply according to published algorithms. Because everyone knows how the Qbit system works, they are free to decide for themselves whether or not they trust it and want to use it.

Not every 2G blockchain currency operates the same way as the Qbit. That's the point. We're about to see an explosion of very cutting-edge currencies hit the market. You, as the consumer, will be able to choose the ones that you like best. If you want to, you can spread your risk over several currencies so that if one collapses, it may hurt but hopefully it won't wipe you out economically.

"The entire human populace is now taking charge of the means of production and changing the rules of the game. They're making their own freaking currencies!" – Paul Vigna

If you are a software developer or an economist (or both), the ability to tailor the rules for currency control is a tremendous power. Anyone who wants to can innovate in this area and find the best methods for keeping the value of currencies stable.

And when it comes right down to it, who do you think is most likely to produce a stable currency: a few people (or even a few thousand) at a government agency or tens of thousands (or tens of millions) of people who each work on their own project that tries to outdo the other and be the best in the market? As for me, I think free market currencies will end up being more stable, valuable, and convenient than anything the government can produce.

---

[115] That's not *strictly* true. In reality the value of the Qbit hovers around a stable average with small ups and downs. Over time, though, the value of the Qbit remains constant.

DECENTRALIZED ASSET EXCHANGES

Second generation blockchains are already making it possible to build distributed asset exchanges. A distributed asset exchange is an online marketplace with no third parties that can control it.

Suppose you have a currency called Woolongs. You want to exchange it for another currency called Wombats[116]. A distributed exchange enables you to place an offer to sell your Woolongs for Wombats at the current market price. Another person can come along and exchange her Wombats for your Woolongs. The entire transaction is handled by the blockchain itself. There are no third parties.

The implications of a distributed exchange are many, but I'll just touch on a few. First, it means that repressive governments cannot ban the currencies you want to exchange. You are free to exchange your Woolongs for Wombats and no one can stop you.

Second, it means that you eliminate what's called *counterparty risk*. When people want to exchange currencies right now, they most go to a third-party money exchanger. That third party may abscond with all your money. With the blockchain, there is no third party. The buyer and the seller both exchange their money with no outside risk imposed by someone who may steal from you.

Third, the stuff that's being exchanged doesn't have to be money. Any digital asset will work. So if you want to build a blockchain system that enables you to sell books, music, movies, and so forth, you can. You can actually make a fortune by doing exactly what Amazon and similar companies do. It can all happen on the blockchain.

THE BLOCKCHAIN ON STEROIDS

Blockchain-based currencies can be much more than just currencies. They can actually provide tools that anyone can use to build new financial instruments, new economic systems, and even new currencies that are specialized for particular uses.

Most people may say, "So what?"

---

[116] Neither of these are a real currency. They were made up by me for this example.

The fact that the blockchain can offer these tools is really quite revolutionary. Imagine a world in which you don't need a bank to get a home loan or a business loan. Instead, you put what's called a *smart contract* into the blockchain and people lend money to you based on standard, fair terms. If your loan is $200,000, you might get one person to lend you the whole amount. Or, more likely, you might get 10,000 people to lend you $20. And it's as easy to pay everyone involved as it is to pay a bank.

Look at it from the other side. How would you like to be able to do what a bank does, just on a smaller scale? Instead of putting your money in the bank for them to invest and make gobs of money on, you can invest it yourself and make that profit for yourself. Essentially, your computer becomes your bank and all the profit the bank used to get comes to you.

How would you like your money stored so securely that it is within the realm of possibility that it could survive a nuclear war? As farfetched as that sounds, it really may be possible with the blockchain[117].

If you could access the same kinds of tools that Wall Street investors use, wouldn't that make it easier for you to invest? The blockchain makes that possible.

What if you could put some solar cells on your roof and sell power to your neighbors easily and automatically? The blockchain may form the basis of a green energy revolution.

And there's more—so very much more. The blockchain literally enables you to create a million killer apps.

The blockchain technology that was pioneered by bitcoin is extensible beyond anything we can imagine right now. A 2G blockchain can contain special data areas, called *metatdata*, that don't exist in the bitcoin blockchain. So it can store information that other 1G blockchains like bitcoin can't.

Here again, this statement sounds boring. But it has a huge potential impact.

What if you could produce an unforgeable ID without the need for any government agency to get involved? There's no reason to have the government issue you an ID if you can do it yourself.

---

[117] I really, really would *not* like to test this theory.

What if you could automate the storage of government records so that anyone can look at them at any time? For example, you could store the complete transaction history of any government agency in a way that is completely transparent to everyone. What would our government be like if anyone could track its expenses and audit it 24 hours a day, 7 days a week?

Imagine that you can issue stock, do an initial public offering (IPO), and provide transparent spending records to investors (and only to them). Now imagine that you can do that without the difficulty or expense of using Wall St. For businesspeople, that's huge.

What if you could create a digital eBay-style marketplace that anyone can sell products into and anyone can become retailers for? And what if that marketplace had no central control so that everyone can compete on an even basis?

Think what you could do if you could bring cloud-based services down to the ground level. How much money could you make if you could offer a digital storage system that enables people to store any information they want in an unhackable form? Now suppose that you don't even have to have a huge data center to make it go. Instead, your cloud storage system just lets anyone add computing resources and disk space to the system and get paid for it. Your cloud storage system could grow to be the largest data storage system on Earth and there would be no central computer that might get hacked or destroyed by an earthquake or other natural disaster.

As you can see, the blockchain offers so many possibilities that it's impossible to list them all here.

INTERTWINED BLOCKCHAINS

Second generation blockchains can also let you create new, compatible blockchains that you can specialize for whatever tasks you want. For example, you can make currencies that are targeted for specific industries.

If you design your specialized currency right, you just might solve America's current health care mess. Really. I'll give a simple example in Chapter 12, but I actually hope you can come up with something better.

Want to create a universal subscription system as a business? Go ahead. You can spin off your own blockchain to do that. A business like that would enable you to sell virtually any service on a subscription basis, including health care.

Could you make money on a system that provided notarized documents (like notary publics do) and permanently stores them for public display?

Or how about a worldwide universal phone system that enables you to make most of your calls for free? And for those calls that aren't free, the system only charges you for a local call? Long distance and international calling fees can become a thing of the past.

Maybe you'd prefer to build a system that provides worldwide Internet access to everyone but is owned by no one? Anyone who wants to can just provide access to everyone else and get paid for it. You can create a blockchain for that too. And you can make money doing it.

Do you want to build an artificially intelligent system to perform a task using information gathered from across the globe? Go right ahead. Create a blockchain that does that.

What if you could create specialized monetary systems that are designed solve massive social problems such as poverty and illiteracy without taxes, government regulations, handouts, or even charity? You can. You just have to build it.

How Real is This?

All of these claims seem pretty fantastic. But the truth is that it is all possible if we just build it. The blockchain provides the foundation and gives us the tools we need to make it happen. So let's move on and see how it all works.

9 THE BLOCKCHAIN IN ACTION

Proposing a free market of competing currencies with custom-designed economies is a bold move. People naturally want to know how such a system works and how they can use it. So this chapter covers that in detail.

Note that come of the concepts covered in this chapter were touched on lightly in previous chapters. But the idea here is to give details so that the advantages of blockchain-based currencies become readily apparent.

PUSH TRANSACTIONS AND PULL TRANSACTIONS

When someone pays you using a blockchain-based digital currency, all you have to give them is an account number to send the money to. They cannot take money out of the account. They only thing they can do is put money in. In other words, blockchain currencies perform *push transactions* because you push money to the seller.

Our credit cards and debit cards work exactly the opposite way. In order to pay with credit and debit cards, you must give people enough information to pull money out of your account. So our current systems use *pull transactions*. The problem with pull transactions is that in order to buy something, you **must** give people enough information to steal from you. That is exactly why we have so much credit and debit card fraud.

> "If I know your address, I can send money to you, but I can't take money out. And that makes a hell of a lot of sense! We wouldn't have the Target security breach if all Target had was an address without any way to use those credit card numbers." — Susan Athey

But because blockchains use push transactions rather than pull transactions, they are **much** more secure than what we currently have. Under blockchain systems, sellers must give people they account information so they can be paid. But people who receive that information do not have any means to steal from anyone else's account. The only way for others to get money from your account is to have you push it to them. If you don't push the money by authorizing the transaction, nothing happens no matter what the other party does. There simply is no way in the blockchain system for them to pull money from you to them. It has to be pushed and it has to be pushed by you. Therefore, using push transactions is more secure and results in less fraud.

In addition, push transactions help protect the free market. The free market is based on the idea that you must consent to pay for something. It requires freedom of choice. But pull transactions, which we currently use, open up the possibility of taking money from you without your consent. Moving to a system that uses only push transactions means everyone must always obtain your permission before they can get any of your money. And that is as it should be in a free country and a free market.

PRIVATE ONLINE TRANSACTIONS

Blockchain currencies guarantee payment at the time transactions are validated. Therefore, the buyer need not be identified in a transaction. This is true even if the transaction occurs online. Shoppers can purchase items in complete privacy without the current mountain of personal data being collected by online retailers.

As with physical cash, the possession of coins is ownership. When the blockchain validates a transaction, the seller receives their money from the buyer. As long as the seller possesses those coins, the seller is paid and has received the value of the transaction just as if physical cash had changed hands.

> Money is a collective agreement. If enough people come to the same agreement, what they agree upon becomes secondary, whether it be farm animals, gold, diamonds, paper, or simply a code. History proves all these cases to be true. Who knows what the future is going suggest to us as money, once we see digital currencies as ordinary?" - S.E. Sever

Figure 9.1 shows how a private online transaction can take place with a blockchain-based currency. As Figure 9.1 demonstrates, a shopper has multiple ways to make private purchases. Typically, the seller will be known to the buyer because the buyer must trust the seller to deliver the goods or services being purchased. The alternative is to use a trusted third party. For example, Amazon and eBay currently serve as trusted third parties for buyers and sellers. Buyers should be aware, however, that dealing with trusted third parties may mean some sacrifice of privacy in order to complete the transaction.

Buyers can preserve their privacy by using digital currencies to pay. Again, sellers are guaranteed payment as soon as the coins are validated. Therefore, the buyer does not need to identify herself. If she is buying a physical object that must be shipped and she seeks even greater privacy, she can have the object shipped to a post office box.

Figure 9.1 A buyer can purchase privately through online or in-person transactions. If the purchased item must be shipped to a buyer, she can increase her privacy by using a P.O. box.

On the other hand, if the purchased item is a download, the buyer can hide her identity by using the Tor network or other privacy software. She is able to reliably pay the seller with untraceable coins and then download her item in complete privacy. Thus she can buy what she wants without having to worry about the intrusive data collection about every aspect of her life that large corporations currently engage in.

PRIVATE IN-PERSON TRANSACTIONS

If a seller can deliver the goods or services at the time of the purchase, which is often the case for in-person purchases, then little or no trust need exist between the buyer and the seller. In such cases, neither the buyer nor the seller need identify themselves to each other. Transactions can be made in complete privacy. Private in-person transactions are nice for people selling things at flea markets and second-hand sales.

In-person transactions can occur in a variety of ways. If the buyer and the seller are both carrying mobile devices (phone, tablet computer, or a laptop) with blockchain-compatible software, no additional parties need be involved. One smartphone can be enough to complete the transaction, though it is more typical for two devices to be involved. Figure 9.2 illustrates how such a transaction takes place.

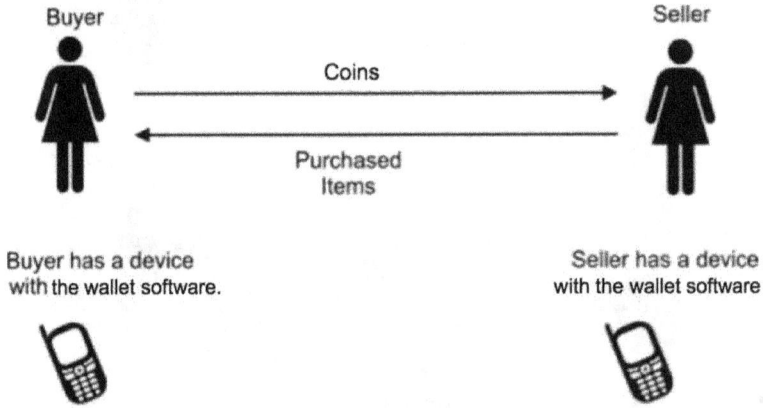

Figure 9.2 Private in-person transactions can occur most easily when both of the participants have a portable device that can validate the transaction.

If both the buyer and the seller have a mobile device, then paying is simply a matter of using a Bluetooth connection, sending a text message, or sending an email from the seller to the buyer. The seller transmits (via Bluetooth, text message, or email) her wallet address to the buyer. The buyer then sends coins to that address.

The seller can also have the wallet software display a 2D bar code on her device's screen. The buyer can use her device's camera to scan the bar code and send the coins.

When only the seller is carrying a device equipped with wallet software, the transaction cannot take place. Because blockchain-based transactions are all push transactions, the buyer must have some way to push her money to the seller. That requires a computer, laptop, or other device.

If the buyer has a mobile device with wallet software and the seller does not, then the buyer can still complete the transaction. The seller can present the buyer with a printed 2D bar code with the seller's wallet address. The buyer can then send the coins to the seller. Later, when the seller returns to her computer, she can sync her computer's wallet software with the blockchain. When she does, she will receive the buyer's coins.

Note

Customers should be aware that governments may require digital cash to be accounted for as part of their citizens' annual tax assessments. The newer wallet software will help you to print all of your receipts so that you can provide an audit trail upon demand.

If two people want to make a digital cash transaction via dumb phones (as opposed to more advanced smartphones such as the iPhone), they can do so if their phone manufacturer has included the wallet software on the phone. This software is usually free and open source, so any phone manufacturer can create a version of it for their phone as an added feature. Currently, very few phone manufacturers have done this. However, you can expect the number of phones with built-in digital wallets to increase as time goes on.

These days, even dumb phones have some kind of data connection. The wallet software uses the data connection to send the coins from the buyer to the seller. Without some form of data connection on the phone, the transaction can't be completed.

When performing a transaction with dumb phones, the seller can transmit the destination wallet address to the buyer using Bluetooth (if it's supported on the phone), a text message, or an email. The buyer runs the wallet software on the phone and sends the payment. The seller's wallet software indicates when the payment is received.

Smartphones are usually a better choice for performing transactions with digital currencies. The digital wallet software will run on many smartphones.

"If we do move toward a digital currency, we should ensure that we retain some type of digital cash that is anonymous — or at least pseudonymous — and that is not tied to a financial intermediary that can control transactions. The nascent digital currency Bitcoin has demonstrated that such e-cash is possible. Bitcoin employs no user-identifiable accounts, relying instead on public key cryptography, so there is no way to know who gave money to whom. And because no intermediaries are needed for Bitcoin transactions, governments have no intermediaries to regulate." — Jerry Brito

As with dumb phones, the seller's wallet software on his smartphone can use text messages to communicate[118] the destination address for the buyer to transmit money to. In addition, wallet software running on smartphones enables the seller to transmit the destination address to the buyer over a Bluetooth, NFC, email, or a Wi-Fi connection.

If the seller has a smartphone, the wallet software can also display the seller's account number on his phone's screen as a 2D bar code. The buyer can use her phone's camera to snap a picture of the bar code and thus send the coins.

Private in-person transactions can also occur at your corner grocery store or any department store. As mentioned in previous examples, these transactions can occur if the seller has the digital wallet or digital cash register software that can process them. The buyer must also have a device with the digital wallet software.

EASY MULTICURRENCY TRANSACTIONS

Shopping in a store where many currencies are accepted may appear daunting at first. However, blockchain systems can be designed to make it as easy as using one national currency.

"When you pay for something with a credit card, you are giving someone all the information they need to know to buy something online. What an idiotic way to architect a system — that if I pay someone, I tell them everything they need to spend my money." – Susan Athey

In the most common scenario, a shopper in a store views an item he would like to buy. On the price tag, he sees the price in his local currency. In addition, there is a bar code that contains the price of the item. The barcode also has information about which currencies the store accepts.

Suppose that a shopper in the US wants to buy a gallon of milk. Imagine further that he has six digital currencies stored in the digital wallet software on his phone. Before going to the store, he selects Wombats, Woolongs, and MegaBucks[119] as the currencies he might use to pay.

At the store, the shopper checks the price shown on the price tag. Of course, it is in US dollars (if he's shopping in the US). But he knows the store accepts multiple currencies. He simply holds up his phone and points its camera at the bar code on the price tag. The digital wallet software uses the information

---

[118] Sending texts may cost money.
[119] None of these are real digital currencies. They are just names that I made up for the sake of this example.

in the bar code to verify that the store accepts two of the three currencies he has chosen to pay with. One of the currencies that the store accepts is Woolongs.

The wallet software displays the price of the gallon of milk in all of the currencies that he wants to pay with and that the store accepts. In this example, he selected three currencies to pay with but the store only accepts two of them. Therefore, the software shows the price of the milk in those two currencies.

The shopper is satisfied with the price of the milk and decides to pay Woolongs for it. He picks up a gallon of milk and goes to the register to pay. Using the wallet software on his phone, the shopper pays for his milk and heads home to enjoy his life.

> "Cryptocurrency Protocols Are Like Onions... One common design philosophy among many cryptocurrency 2.0 protocols is the idea that, just like the internet, cryptocurrency design would work best if protocols split off into different layers. Under this strain of thought, Bitcoin is to be thought of as a sort of TCP/IP of the cryptocurrency ecosystem, and other next-generation protocols can be built on top of Bitcoin much like we have SMTP for email, HTTP for webpages and XMPP for chat all on top of TCP as a common underlying data layer."
> Vitalik Buterin

STORE YOUR MONEY PRIVATELY AND SAFELY

Blockchain-based currency systems can provide free wallet software for people to put on their handheld devices and personal computers so that they can store their coins. These systems can also offers digital safe software (like a bank safe or a safe hidden in your wall) that backs up customers' wallets. It can store the coins on any digital device that contains writable storage, such as those in the following list.

- Hard drives
- CDs
- DVDs
- USB flash drives
- Cloud storage services (Skydrive, DropBox, etc)
- Email
- Many more.

Coins can even be backed up to paper and put into safe deposit boxes or fireproof safes.

Of course, digital currency systems can be used by banks as well. So it is entirely possible for people to store their coins in banks, just like any other form of currency. Once they do, they can obtain credit and debit cards that are in the digital currencies rather than a national currency.

OBTAINING DIGITAL CASH

To get first generation digital currencies like bitcoin, you can buy them on an online coin exchange. Or you can use an ATM, such as the new bitcoin ATMs that are becoming popular.

Be aware, however, that governments place strict regulations on such businesses. For example, you must disclose your identity to both online exchanges and bitcoin ATMs. In fact, they'll generally ask for a copy of your driver's license and may even scan your handprint.

> "I understand the political ramifications of [blockchains] and I think that government should stay out of them and they should be perfectly legal."
> - Ron Paul

For the newer, 2G currencies that are coming online, you can obtain digital cash by downloading the wallet software and buying them on the system's decentralized, built-in currency exchange. You can also use the wallet software to sell that you own through the decentralized exchange.

In addition, you can sell goods or services in exchange for digital cash. People pay you by sending coins directly from their wallet to yours.

Those who so desire may be able to make arrangements with their employers to be paid in one or more digital currencies.

RESHAPING THE ECONOMY

In the last chapter, we saw how easy digital currencies are to use. This chapter presents scenarios that are difficult or impossible under the current currency systems. This list is not exhaustive. It is just intended to illustrate the power and flexibility of digital currencies based on the blockchain.

ENFRANCHISING DEVELOPING NATIONS IN THE GLOBAL MARKET

People in rural areas of developing nations typically do not have the wherewithal to participate in the global economy. However, many of them have mobile phones. Digital wallet software can be made to run on nearly any mobile phone available. The source code can be made freely available for anyone to adapt to new hardware. Therefore, any phone manufacturer anywhere in the world can include the wallet software on the phones they sell as a free added feature. With the wallet software, people of very limited means can have greater access to local and global markets.

Helping the Unbanked

It's important to understand that producers in developing nations often do not have access to banking and credit card services. This limits their access to many markets—including developed urban markets within their own country.

Imagine, for instance, that a company in India creates a retailer web site enabling rural craftspeople to sell their wares in cities all around India. Such a service would be critical for the prosperity of many in rural settings worldwide.

> "The relative success of the bitcoin proves that money first and foremost depends on trust. Neither gold nor bonds are needed to back up a currency." - Arnon Grunberg

As it stands, it can be difficult to create such a service in some areas of the world. The retail company must take credit cards from customers. It can transmit orders to sellers via text messages. However, the price of purchases is often so low that either most credit card companies won't process the transaction. Or the transaction fees charged by the credit card companies make such purchases unprofitable.

Digital currencies make small transactions possible. Because coins can be sent to even the dumbest of dumb phones, it is possible for merchants and craftspeople to participate in the modern economy with nothing more than a mobile phone.

Entering the Global Marketplace

Suppose, for instance, that a woman making decorated clay pots in her village in India has a phone that does text messaging and has the wallet software on it. She can sell her pots through the retailer web site. The fact that her pot sells for a low price is not a deterrent to processing the transaction.

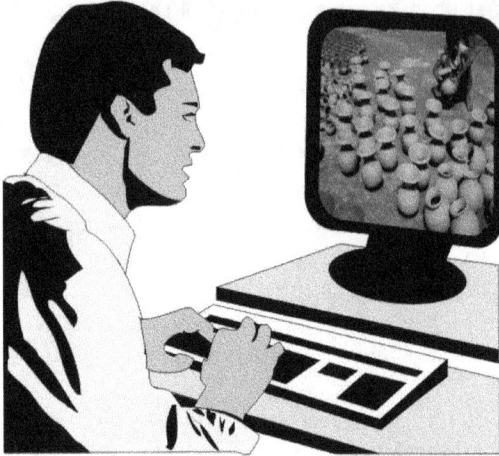

Figure 10.1 A resident of Mumbai buys a pot from a rural village India. Because banks and credit cards are not needed for transactions, those who don't have access to them can still participate in the modern economy.

Imagine that the woman's pots are seen on the web site of the retailer by a man in Mumbai. He appreciates the seller's craftsmanship and artistic style. He decides to buy her pot. He notices that the cost of shipping is actually higher than the price of the pot. But this does not bother him because both the price and the shipping together amount to a reasonable cost in his opinion. He therefore buys the pot.

> "The first generation of the digital revolution brought us the Internet of information. The second generation — powered by blockchain technology — is bringing us the Internet of value: a new platform to reshape the world of business and transform the old order of human affairs for the better." – Don Tapscott

Seeing that the retailer's web site only takes digital cash, the buyer pays for the pot using coins from the digital wallet software on his computer. He inputs the shipping information and completes the transaction. His coins are immediately validated by the blockchain. The retailer sends the order to the potter by text message.

Shipping the pot clear to Mumbai is not a problem for the potter because the retailer sends her payment in digital cash. Her wallet software automatically receives and stores the coins. Of course, the retailer has taken its share of the purchase price. But this does not matter to the potter. By selling her pots online to buyers in faraway cities, she is able to expand her business far beyond her local market. She feels she is getting a good deal.

At her local post office, she obtains the necessary shipping materials (box, padding, and so forth) and sends the pot using digital cash. Alternatively, she can convert the coins to her local currency and pay with that.

The potter sends a text message to the retailer verifying to the retailer that the pot was sent. The retailer then forwards the message to the buyer.

Meanwhile, the pot arrives at the Mumbai flat that is its destination. The buyer finds it everything he was hoping it would be and happily displays it in his home. Like the potter, he also feels he got a good deal.

This scenario works because:

- The transaction fees are extremely low.

- The buyer can purchase digital cash in his local currency.

- The buyer can order the pot using digital cash.

Keeping Coins Safe

To keep her coins safe, the potter forwards them to a free digital safe provider using the web browser on her phone. Of course, she must put up with the ads on the digital safe provider's web pages. But those ads make the digital safe service free to her.

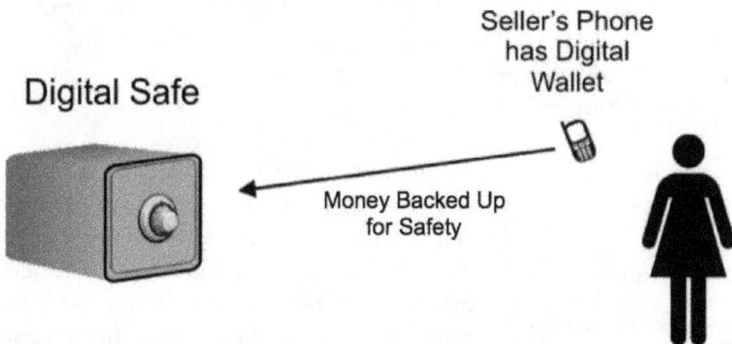

Figure 10.2 Online services can provide a digital equivalent of a safe so that the potter's coins are always backed up. Even if her phone dies, she won't lose her money.

The digital safe software makes an encrypted backup of the potter's wallet. If her phone is lost, stolen, or gets broken, she can restore her wallet to her new phone without losing her money.

INCENTIVIZING PURCHASES

There are many ways to get people to purchase goods or services. One of them is to incentivize the purchase by offering some sort of rebate or cash back program. Many merchants have such programs. When closely examined, it can be seen that these are generally a form of digital currency or a digital substitute for currency.

Digital Purchases

When consumers buy digital products online, sellers can use digital currencies to incentivize purchases. For example, a small publisher of eBooks can include a bar code at the end of the book that contains some coins from a special-purpose blockchain. These coins can be used to get discounts on future purchases, thus providing an additional incentive for the reader to buy the book. The purchaser of the eBook needs only to use his digital wallet to access and save the coins. He can then spend them exactly one time. Once the coins are procured from the eBook, they are no longer in the book[120].

What's happening in this example is that the merchant is using a digital currency to issue merchant-specific discount cash. Merchant-specific currencies can only be redeemed by submitting it to the issuing merchant. This is an easy and inexpensive way for merchants to give store credit to purchasers every time they buy something.

> "Bitcoin gives us, for the first time, a way for one Internet user to transfer a unique piece of digital property to another Internet user, such that the transfer is guaranteed to be safe and secure, everyone knows that the transfer has taken place, and nobody can challenge the legitimacy of the transfer. The consequences of this breakthrough are hard to overstate."
> Marc Andreesen

As an illustration of this, suppose the publishing company is called East End Publishers. They can name their merchant-specific currency East End Bucks. The East End Bucks can be stored in the customer's wallet. When the customer wants to buy another eBook from East End Publishers, he can use the East End Bucks as partial (or even complete) payment.

---

[120] I am *not* indicating in this scenario that a pull transaction is being performed by the buyer when the buyer retrieves the coins. The bar code in the book would give direct the buyer's device to a web service that provided the coins using a push transaction.

Likewise, sellers of apps for mobile devices such as Android phones can incentivize the purchase of their apps by adding a one-time function to the app that gives the buyer some of their own branded digital currency. Many other types of digital products can be incentivized with digital currencies.

The Return of Green Stamps

If you're old enough, you may remember the days when your mother received Green Stamps with her purchases. Green Stamps, for those who don't know, were actual stamps printed by the Green Stamp company. Merchants bought Green Stamps and then assigned a certain number of them to the products they wanted to promote. That is, you might go into a store and buy toothpaste. When you paid for your toothpaste, the cashier might give you five Green Stamps. You then took the Green Stamps home and pasted them into Green Stamp books (they had adhesive on the back, just like postage stamps). When you filled enough books, you could order something from the Green Stamp company's catalog and send the books to pay for the item.

> "The blockchain ... facilitates new types of economic organization and governance. [It] suggests two approaches to economics of blockchain: innovation-centred and governance-centred. [It] argues that the governance approach—based in new institutional economics and public choice economics—is most promising, because it models blockchain as a new technology for creating spontaneous organizations, i.e. new types of economies." — Primavera De Filippi

These days, merchants can do the same thing with digital cash. The large grocery store chains, for instance, offer "cash back" or "discount" cards. Both these and Green Stamps are different forms of private currencies.

At this time, small merchants can't compete. A mom-and-pop grocery store doesn't have the wherewithal to issue member discount or cash back cards. However, digital currencies enable them to give digital rewards just like the big stores.

With digital currencies, a small merchant can purchase Green Stamps as digital coins from a company that provides the service. When the customer makes a purchase at the merchant's store, the merchant can send some Green Stamps to the customer's wallet as a reward. Eventually, the customer can use the Green Stamps to order something from the Green Stamp provider's catalog.

> "The first generation of the digital revolution brought us the Internet of information. The second generation — powered by blockchain technology — is bringing us the Internet of value: a new platform to reshape the world of business and transform the old order of human affairs for the better." — Don Tapscott

By using a service like this, merchants can incentivize their customer's purchases. Through collecting Green Stamp coins, the customer is rewarded by being able to buy something from the Green Stamp company's online catalog. Of course, this catalog gives the customer access to products that the local retailer is unlikely to have. And the Green Stamp provider makes money when local retailers buy their Green Stamps.

AD HOC RESOURCE CONTRIBUTIONS

Digital currencies can enable owners of computers to make money by contributing idle computer time to processing networks in an ad-hoc way. For instance, suppose a trusted company called Big Time Transactions publishes a software program called Transact. Imagine that Transact does some sort of business transaction processing. Anyone can download Transact for free. People are willing do this because they make money and because they trust Big Time Transactions and know that Big Time Transactions will not try to hack their computers.

Imagine that Mary downloads Transact. She can leave her computer on when she is not using it. The Transact software automatically goes to the Big Time Transactions' server, gets some data, and processes it.

"Instant transactions, no waiting for checks to clear, no chargebacks (merchants will like this), no account freezes (look out Paypal), no international wire transfer fee, no fees of any kind, no minimum balance, no maximum balance, worldwide access, always open, no waiting for business hours to make transactions, no waiting for an account to be approved before transacting, open an account in a few seconds, as easy as email, no bank account needed, extremely poor people can use it, extremely wealthy people can use it, no printing press, no hyper-inflation, no debt limit votes, no bank bailouts, completely voluntary. This sounds like the best payment system in the world!"
Trace Mayer

Mary gets a small fraction of a coin for each batch of data that her computer processes. Throughout the day, these add up to enough coins to provide a tidy profit for Mary. The Transact software deposits the coins into her digital wallet in exchange for the use of her computer while she is away. This turns idle processing power into a source of revenue.

Mary does not have to give Big Time Transactions any personal information at all in order to receive her money. Big Time Transactions does not know or care who contributes computing resources to the network. Mary's privacy is protected. Because Mary trusts Big Time Transactions, she can be assured that she will be paid each time her computer processes a batch of data.

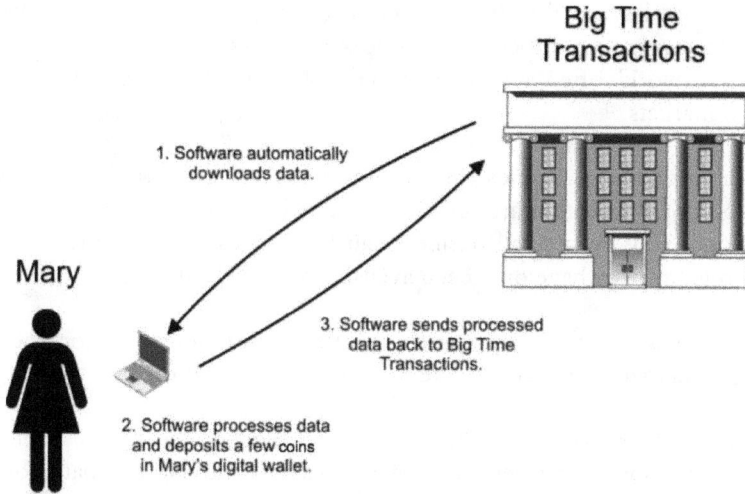

Big Time Transactions

Mary

1. Software automatically downloads data.

3. Software sends processed data back to Big Time Transactions.

2. Software processes data and deposits a few coins in Mary's digital wallet.

Figure 9.7 Each time Mary's computer is not in use, it can make some money for Mary by "renting out" its processing power.

In addition, ad hoc resource contributions also enable companies to do large amounts of processing without excessive capital investments in computer hardware. While this is not an issue for large businesses, it is critical for small to medium-sized companies. Distributed systems like the one presented in this example can give smaller enterprises the ability to go head-to-head with massive corporations.

Under the current monetary system, a scenario such as this is nearly impossible. The transaction fees would be too high to enable microtransactions like these.

THE NEW ECONOMY

Digital cash enables us to build a new economy. With it, you can perform completely private transactions, both online and in person. It enables anyone to use multiple currencies as easily as they use just one. Digital currencies don't need to be stored in banks. You can store them or back them up on multiple personal devices that are under your control.

Digital currencies are able to enfranchise many people who are unable to access the transaction processing services of banks. These unbanked poor around the world often remain in poverty because they have little or no access to markets.

In addition, digital currencies can be used to incentivize purchases through special-purpose coins that are only redeemable by a specific vendor or group of vendors. Now even small business can have access to the kind of "cash-back" programs that have only been available to large businesses.

With digital cash, any company or organization can build an ad-hoc network of computers that people contribute in order to make money.

These and other scenarios show how the blockchain can reshape the economy into something new, something that will provide real paths out of poverty.

But digital currencies are just the beginning. The blockchain can provide additional tools that will help set off a prosperity revolution. And while I realize that that is a huge claim, it is nonetheless true. The remaining chapters in this book present an overview of what these powerful tools are and how they can reshape our economy, our government, and our lives.

PART 3

ANATOMY OF A PROSPERITY REVOLUTION

10 SMART CONTRACTS: THE REBIRTH OF CIVILIZATION

Unlike the bitcoin blockchain, 2G blockchains can be designed to store specific types of data, called *metadata*, that enable software developers to build additional functionality into the blockchain economy that no one has thought of yet.

In other words, by storing metadata in the blockchain, we can enhance the blockchain's functionality and build entire economic infrastructures on top of it. The blockchain essentially becomes a "smart" entity.

The most important innovation that the blockchain's metadata enables us to build is called a *smart contract*. Smart contracts are a feature of the blockchain that will completely turn the world upside down. It's the feature that enables us to build a billion killer apps on top of the blockchain. Seriously.

So let's find out what they are.

WHAT IS A SMART CONTRACT?

A smart contract is a legal contract just like any other legal document. It contains all of the normal language (well, a *lawyer's* version of normal language) that a legal contract contains.

In addition, smart contracts contain scripting code that is executable by the blockchain system. The scripting code executes the terms of the contract. So a contract is both a legal document and a computer program that does whatever the English legal clauses say.

Why is that a big deal?

NEW INFRASTRUCTURE

Smart contracts enable anyone to create new economic infrastructure based on the blockchain and that infrastructure functions just as if it was built into the blockchain system. Specifically, smart contracts:

1. Bring the abilities that are normally only available to large corporations down to the level of the independent investor.

2. Enable anyone to used "crowdsourced" financing for just about anything.

3. Automate the negotiations of transactions and thereby let us create new ways to make money that were never possible before.

Let's look at each of these in turn. But please understand that we're not going to delve too deeply into any of them at this point. We'll start with an overview and get the detail shortly, so let's just look at some simple examples of the possibilities presented by smart contracts. After that, we'll examine how smart contracts work. Then we can dive into some examples of smart contracts that are more detailed and that demonstrate the ideas presented here.

Banking without the Bank

Suppose you get a mortgage that the bank implements as a smart contract. The contract contains a legal statement in English saying that you have to make your payment before the 5th of the month. The program code that corresponds to that statement manages a blockchain account that functions just like a bank account. If your payment is not in that blockchain account by the 5th, the contract automatically sends you an email telling you that your payment is late. It may even make an automated phone call to your home phone to tell you that you'll be assessed a late fee.

Let's say that you make your payment on time. The program code in the smart contract sees that you've made your payment into the account that it manages. It sends a notification to the bank's computer so that the bank's computer automatically records your payment. It also emails you a receipt for your payment.

Smart contracts can manage money, pay dividends to investors, hold money in escrow, and do many other functions that are required for processing financial transactions. In fact, they can do most of what a bank can do without the bank.

You can also use smart contracts to create new types of investments that we do not have now. With smart contracts, you can build any kind of economic infrastructure that you can think of.

Microfinancing

Did you know that a bee makes 1/12 of one teaspoon of honey in its entire lifetime? It takes two million bees working together to make one pound of honey[121].

That's interesting, but what does it have to do with smart contracts and the blockchain?

Smart contracts enable lots of small investors to contribute whatever they can to a business endeavor of any kind. When the business endeavor earns profits, those profits come back to the smart contract. The contract then pays each investor in direct proportion to their investment.

What this means is that anyone and everyone can become a micro-scale venture capitalist and reap the rewards of it. Anyone and everyone can invest their small amounts of money into home loans, college loans, car loans, business loans, and more. They do this by simply signing smart contracts and depositing their digital cash in escrow into those contracts. The contracts do everything else.

Automated Business Transactions

A lot of people know that Wall Street firms do most of their trading with automated computer programs. But smart contracts extend that capability to everyone.

In addition, smart contracts can negotiate for the use of resources that people make available so that they can earn a profit. These might be computing resources, Internet access, or a whole host of other services.

You might be walking down the street and need quick, temporary Internet access. A private citizen in an apartment overhead might have an Internet access point that she has put up specifically to sell access to people on the street. Your mobile device can use a smart contract to negotiate access to her Internet access point in return for a small amount of money (probably less

---

[121] http://www.honey.com/newsroom/press-kits/honey-trivia

than 1 cent). It all happens invisibly to you. All you see is that you get the Internet access that you need.

HOW DO SMART CONTRACTS WORK?

Smart contracts are computer-based protocols that facilitate or enforce the conditions of a legal contract.

The simplest example of a smart contract scripting code is just an IF and a THEN, with possibly an ELSE. Figure 10.1 illustrates an IF-THEN-ELSE statement in a smart contract.

```
IF (some condition is TRUE) THEN
        Do a task.
        Another task.
        Still another task.
ELSE
        Do an alternate task.
        Another alternate task.
        Yet another alternate task.
END-IF
```

Figure 10.1 A Simple IF-THEN-ELSE in a
Smart Contract

Anyone who has ever programmed a computer will recognize this sort of statement. It says that if the condition, which is contained in the parentheses, is true, then it will execute the tasks between the THEN and the ELSE. However, if the condition is not true, it executes the statements between the ELSE and the END-IF.

Real smart contracts are far more complex than this but the statement in Figure 10.1 illustrates the basic building blocks of a smart contract. And real legal contracts can be based on this concept. Figure 10.2 shows some examples.

```
IF (my partner quits) THEN
        The money will be divided as follows.
ELSE
        We divide the money this way.
END-IF


IF (Author A writes 25% of a book) THEN
        Pay Author A 25% of the royalties.
ELSE
        Author A makes nothing on the sales of the book.
END-IF
```

Figure 11.1 IF-THEN-ELSE Examples from
Real World Contracts

In an IF-THEN-ELSE statement in a smart contract, the condition is called the *oracle*. The oracle is a trusted data source that validates the conditions of the contract.

For example, if I deposit my mortgage payment into an account managed by a smart contract, then the smart contract can see that. Therefore, the oracle is the account. Either I put the money into it or I don't. Because the account is a reliable indicator of whether or not I'm complying, the contract can use that as a source of information that is absolutely guaranteed to be right.

> "You can call it programmable money: It's money that we can write computer programs on, and these computer programs check when certain conditions are met." – Susan Athey

As previously mentioned, smart contracts can hold money in escrow so that it isn't available to any party in the contract until the oracle says it's ok to pay it out. Therefore, no third party is needed to guarantee payment. This is actually quite revolutionary.

CAN I MAKE MONEY WRITING SMART CONTRACTS?

Building smart contracts is an incredible business opportunity. Let's suppose that you want to create a business that produces standard smart contracts such as rental agreements, agreements for selling homes, and so forth. You can post your contracts on your web site for free and sign them in a way that makes specific clauses unchangeable. Anyone can download your contracts. But your contracts can all include an unchangeable clause which states that

> "Blockchain verifies the information using the following steps: Consensus - it requires the majority of the block builders to agree that the occurrence actually happened. Consistency - requires that the new information fits with the previous block. Transaction - it requires that the transaction occurred by looking at the previous block, ensuring that two people did not record conflicting accounts of the information. Automated Conflict Identifiers - the software itself trolls for conflicts within the blocks and the structure. There is no centralized location, or big computer in the sky, where the information can be altered or stolen." — Jacob William

you must be paid for each use of your contract. So anyone can download your contract and each time they use it for a new agreement, the contract automatically pays you. If the people using your contract don't pay, your contract refuses to execute. It becomes useless to them. In this way, the contracts that you produce automatically insure that you get paid each time someone copies and uses them.

Here again, you don't need a third party to sell your contracts for you. You can simply post them on your web site and know that any time anyone uses a contract you've created, you'll get paid. More than likely, the amount you get paid will be small. But like millions of bees contributing a small amount of honey to the hive, many people using your contracts will contribute to a healthy cash flow to you.

SMART CONTRACT EXAMPLES

So let's take a look at some more in-depth examples of how we could use smart contracts. Some of these examples could be implemented right away. Others would take some development and investment. But I'm using them to show how we could literally remake our civilization if we are willing to rethink our money and our economy.

EXAMPLE 1: MOBILE WI-FI AND PHONE ACCESS

Let's suppose that you are riding in your car and you need to use your iPad to get driving directions. Someone else is driving, so you don't hesitate to pull out your iPad and look up the address of your destination.

Now imagine that your iPad is one of the less expensive ones that doesn't have a connection to a phone network. It only has Wi-Fi. So what you need is instant access to a cheap Wi-Fi connection. That's easy to provide.

Cheap Wi-Fi through Smart Contracts

Suppose that the car you are in is at a red light near a house that has a special Wi-Fi device[122]. This device gives anyone safe access to the homeowner's Internet connection. You have some digital cash on your iPad. With the right app, your iPad can automatically negotiate with the homeowner's Wi-Fi device for Internet access while you are in range—even if you're only in range for a few seconds. Your iPad gives the homeowner's device a small fraction of a coin and the homeowner's device gives you Wi-Fi access. You look up the address of your destination.

Now imagine that the light turns green and the car rolls on down the street. As you go, the app on your iPad keeps finding Wi-Fi hotspots put up by the owners of each building along the street. It connects with one after another as you move along the road. Each time, it pays for a brief amount of access (usually lasting from a few seconds to a few fractions of one second). It seamlessly keeps you connected as you continue down the road[123].

This scenario is possible because apps can use smart contracts to negotiate for Internet access as you drive. The Internet access app places your digital cash in escrow and pays when you receives access. No third party or human intervention is needed. It just uses a standard contract that both devices recognize and trust.

MANETs and the Death of the ISP

Providing Internet access in this way is called *mobile ad hoc networking* (MANET). MANETs, when combined with digital currencies and smart contracts, can provide everyone with universal access to the Internet for far less money than we spend now. Everyone who wants to can put a MANET access point in any building that they own and make money. With a device that plugs into your car's electrical outlet, it is even possible to pass network traffic from car to car on a highway so that the highway becomes a data conduit. MANETs essentially turn everyone into Internet service providers (ISPs). And in doing so, they democratize Internet access in a way that is currently not possible.

---

[122] This device is actually just a standard Wi-Fi router connected to a boosted antenna, both of which can currently be purchased on the market from a number of sources. The router would require special software to make this example work.

[123] This is not hard technology to develop. We already use algorithms like this for moving your phone connection from cell tower to cell tower as you drive or walk along a street.

EXAMPLE 2: TURNING ROADS INTO MONEY SOURCES

With smart contracts, many revenue "sinks" can become revenue sources. A revenue sink is something that drains your money away. A revenue source produces income for you.

For instance, right now, it takes a lot of money to build and maintain roads. They are revenue sinks. What if roads paid for themselves?

Roads as Communications Conduits

This example builds on the ideas presented in Example 1. Let's imagine that a rural county decided to defray the expenses involved with maintaining their highways by providing a rectangular concrete base that sits on the shoulder at regular intervals along each side of the highway. These concrete bases are just rectangular slabs of concrete set at regular intervals, say one mile, along the highways.

Suppose further that these highways are fairly heavily traveled because the rural area is between to heavily populated cities. The county opens up bidding on the lease for each individual concrete base. The highest bidder gets the right to lease the base. The leaseholder can then mount a device, called a *base station*, onto the concrete slab they have leased.

The base station on the concrete slab provides Wi-Fi access to passing cars. It also passes all of the network traffic it handles on to the next base station on the next concrete pad along the highway. It can pass network traffic in either direction along the highway.

In this scenario, every base station along the highway can make money from the passing cars. They provide Internet connectivity to cars while the cars are in range. They then pass the network packets up and down the highway until the network packets get to a base station that has a broadband Internet connection.

Imaging that every ten miles or so, there's a base station with broadband Internet access. Or it might be every fifty miles, or whatever else is appropriate. Of course, the base stations with broadband access can command a higher lease because all of the other base stations along the highway pass network traffic to them. But either way, the vehicle traffic along the highway pays the owners of the base stations for access. And the money from the leases helps pay for the maintenance of the highway.

Now let's extend this example and say that the base stations along the highway also provide broadband access to all of the buildings within range of the highway. This is another source of money for the people who provide the base stations.

We can go even further if we recall that this highway connects two large cities. And let's say that both of these two cities have MANETs in them. The county can install base stations at the edges of the cities that enable network traffic from the two MANETs to be passed along the highway. So network traffic from the two cities is moving back and forth along the highway. That's yet another source of income for the county to spend on its roads. All of our highways can become Internet and telecommunications corridors.

Reimagining What a Road Is

But let's push this idea even further. Let's imagine a new way of building roads. Suppose that we build about two dozen conduits running through the road's base? Let's say that half of them could be used for power lines. The other half were designed for fiberoptic networking cables. At regular intervals, there are manholes that give access to the conduits under the road.

Now what good does this do?

Well first, it enables power companies to come in and run cables under the road relatively easily. They just pull their cables through the conduits. That's nice, but not revolutionary.

Building roads in this way enables competing power companies to service the same area because multiple power companies can pull cables through the roadways. You could have twelve power companies that each pulled a cable through their own conduit in the road. They would then compete to sell power to all of the consumers along the roadway. Gone would be the days of government-granted power monopolies. Multiple competing power companies would drive down the price of electricity and drive up the quality of service. The government would lease the power conduits under the road to the highest bidder and use the funds to maintain the road.

And because it is possible to send network traffic across power lines, the power companies could also become ISPs and provide Internet access to anyone along the road[124].

Recall that in this scenario there are also a dozen conduits under the road for fiberoptic cables. Telecommunications companies and ISPs could bid on the leases for these conduits and also provide faster, more expensive Internet access.

Smart Roads and the Free Market

To really make this a mind-blowing scenario, let's say that each section of the road is owned by a different leaseholder. In other words, if I'm an ISP with a fiberoptic cable in one section of the road, my cable[125] can automatically negotiate with all twelve of the cables in the next road section to find the lowest price for passing along network traffic.

The power cables can do the same with electricity. That is, they negotiate with the cables at either end of their section of the roadway to buy power at the lowest price and sell it to consumers in buildings at the highest price they can get. But of course, the buildings have all have computers with smart contracts too. They buildings are automatically negotiating with the twelve power cables in the roadway for the lowest price on electricity.

Literally thousands of small companies may own cables installed along a lengthy section of a roadway or under the streets of any given city. Each cable makes money and each cable negotiates for the best price for passing along its network traffic or electrical power.

Digital currencies and smart contracts make this all possible.

It's even possible that government would have no more involvement in roads other than to plan them and buy the land for them. They could then lease the land to companies that build and maintain roads. Those companies would build and maintain the roads at their own expense. They would make money by leasing the conduits under the road to the highest bidder.

---

[124] Internet access via power line connections are slower than fiberoptic cables. The power companies would necessarily be competing for customers that want cheaper access and are satisfied with slower speeds. Even so, the speeds available to customers through power lines are comparable to what most people have today. In other words, the competition would most likely give them the same speeds they have now but at a lower price.
[125] Fiberoptic and power cables usually have tiny computers installed in them at regular intervals.

Smart Roads Pave the Way to Clean Energy

Scenarios like these also make it possible and profitable for us to move to alternate sources of energy that are not now profitable. Currently, the government heavily regulates the production and sale of electrical power. And the power companies are powerful[126] government-granted monopolies[127] that use their status to keep things regulated in their favor. The current fee structures which are imposed by these onerous regulations make it nearly impossible for small electrical producers to exist.

But what if our roads enabled us to connect to any one of a dozen power companies? And what if there were virtually no regulations governing the production and sale of power? In such an environment, small solar and wind producers could thrive.

> "The power company of the future, many experts say, will feature new electricity rate structures that reward efficiency, finance and integrate local, on-site power generation (like rooftop solar)." — Dick Munson

If the system I'm proposing existed, all along the southern section of the US, property owners could easily install solar panels to produce electricity and sell it to the highest bidding power cable in the roadway that connects to their property. For property owners, it would simply be a matter of installing the right equipment and plugging it into the road. The rest would be automatic.

If the system I'm proposing existed, most of the northern US could be a source of wind-generated power. Again, it would simply be a matter of plugging the right equipment into the road. The power would be sold to the highest bidder and the property owner would make money.

While it's unlikely that property owners would get rich this way, the opportunity to easily make a profit would create a boom in wind and solar power. By enabling innovation that is impossible now due to the heavy hand of government control, we could profoundly impact our environment for the better *and* make money doing it. This would also make our power grid much more resilient and more resistant to natural disasters and terrorist attacks.

But you may rightly ask, "What about times when there is no sun or wind, such as on a calm summer night?"

Good question.

---

[126] Ok, yes. Very punny.

[127] Government-granted, centralized electrical power monopolies were invented nearly 100 years ago and little in the industry has changed since. And that is precisely because government regulations have locked out innovation and change.

The answer might be shared batteries. Suppose I am a retired navy engineer who has worked on the giant batteries used in nuclear submarines[128]. Imagine that I take my life savings and buy a disused building in a city. Inside the building, I place several of the giant batteries that nuclear submarines use. I then connect a computer to them and connect them to the power grid in the street.

During the day, when there is cheap power from solar and wind generators available, my computer automatically buys electricity from the grid and recharges my giant batteries. During the night, when people are home watching TV, taking showers, and doing other things that require electricity, my computer does the opposite. It sells power stored in my batteries back to the electrical grid at a higher price. Smart contracts and digital currencies make this relatively straightforward to implement.

In this way, our entire green power structure would assemble itself. It would become a distributed, decentralized, self-organizing system—a DDSOS. Remember those? A DDSOS makes it possible for us to build large, complex, and highly resilient power systems with little or no government intervention. The system assembles itself using green energy sources simply because people want to make a profit.

Our current government regulations prevent any of this from happening. And I will even go so far as to assert that without the ability to use digital cash, smart contracts, and a DDSOS, we will *never* see a green energy revolution no matter how much tax money we dump into it.

EXAMPLE 3: MICROLENDING AND MICROINVESTING

What if everyone could become a lender and give the banks competition?

That's not an idle question in the world of digital currencies and smart contracts.

Suppose that I want a mortgage to buy my house. Digital currencies and smart contracts can alter this process and make it something that is more democratic and more profitable than it is now.

---

[128] Nuclear submarines use their nuclear power plant to charge huge batteries that are often bigger than the average garden shed. Special care must be taken around these batteries because of the electrical fields they emit. For example, you can't have anything metal on your person when you get near them or you could be seriously injured. Working on these batteries safely requires special knowledge and training.

Right now, if you want to buy a house you go to a bank to get your loan. Because the banks have no real competition, they can dictate the terms of the loan to be *very* favorable to themselves. But digital currencies and smart contracts can change that.

Also, as it currently stands, you can only be a lender if you have a lot of money to lend. This limits the possible investments available to small investors. Digital currencies and smart contracts can democratize this situation by making it possible for even very small investors to participate in the lending process. Let's see how it works.

Democratizing Lending

Ok, so you're buying a house. Instead of going to a bank, you go to a private investment group for a 15-year mortgage. The investment group vets you just like the bank does. That is, they look at your financial situation like a bank does to determine whether or not you're a good risk.

Imagine that the investment group sees you as a good risk. Instead of having a big pile of cash on hand to lend you, the investment group posts a smart contract on its server. Anyone in the group can look at the contract and determine whether or not they want to invest. Suppose my grandmother in Poughkeepsie, New York has just $10 USD to invest every month. She has invested with this group before, so she trusts them.

Granny signs the contract and invests her $10. So do 15,000 other people. Everyone invests just $10. Now you have the $150,000 you need to buy your house. You make your monthly payments to the contract and the contract automatically distributes a small portion of your payment to every one of the investors, including Granny.

Neither Granny nor any of the other investors is going to make much money on that investment. Over the life of the 15-year mortgage, Granny is probably going to make only about $25-$30. Doesn't sound like much, does it? But let's say that Granny has been investing $10 per month this way since my Dad was in diapers. All of that adds up. Suddenly, Granny has a very comfortable retirement.

Democratizing Insurance

Now you're looking for homeowner's insurance. Well guess what? The investment group also offers that. And my Uncle Marlin, who's also a member of the investment group, sees the smart contract for your insurance policy. He decides to invest. His $10 sits in an escrow account controlled by the contract. No one can do anything with that money until the policy is either paid out or cancelled. You are guaranteed that your money is there when you need it. Until then, a small portion of each of your monthly payments will go to my Uncle.

If something happens, such as a fire, there is a payout. But Uncle Marlin has only invested $10 in your policy, so it's not a great loss. He can spread his risk across many insurance policies and probably make money with the majority of them.

Or you can take the opposite approach to microinvesting. If you need $150,000 of homeowner's insurance, you can post 150,000 smart contracts on the Internet that are each worth $1.

What good is that?

Many people have had the experience of trying to file an insurance claim and having the insurance company find a technicality that lets them not pay. They have a vested financial interest in not paying you that is large enough for them to hire lawyers to defend themselves against you.

Instead of doing that, you could have 150,000 insurers that insure for $1. Most of them would pay because the payout is small for each insurer. Some might not, but your loss wouldn't be significant because none of your insurers have a large enough investment to want to defend themselves in court for non-payment if you sue them. In other words, microinvesting can lead to fairer agreements.

The point here is that microlending, microinvesting, and microinsurance are possible and profitable with digital currencies and smart contracts. They enable a more democratic lending structure because far more people can participate in the lending process. And microlending means that far more investments are available to the average person, so the money that is currently concentrated in the hands of big bankers gets spread out across the economy.

Microlending and Venture Capital

Microlending can also profoundly affect the venture capital industry. Currently, venture capitalists are big money players who stack the deck in their own favor. Microlending enables *anyone* to invest in new companies through smart contracts. Again, you can spread your risk over many investments so that the ones that go south don't impact you that much.

But microlending means that more people will be investing in the venture capital space. More investors means that it will be easier for small companies to get the startup funds which they need to succeed. Right now, that's excruciatingly difficult for them. This one change can have a profoundly positive effect on the small business sector of our economy, which has been the hardest hit since the 2008-2009 meltdown.

In addition, microlending and microinvesting through smart contracts also means that lenders can now be located all over the world. We no longer have to be limited to investing in our own country. If we feel that a new service in a developing country is a worthwhile investment, we can invest our money there and help advance countries that are currently struggling to modernize.

Digital currencies and smart contracts also make automatic reputation systems possible. An automatic reputation system would enable more people across the world to do business with each other because the reputations of the buyer, lender, or insurer would follow them into future transactions. This vastly lowers the cost and the risk of doing business and rewards those who are trustworthy while pushing the flakes, frauds, thieves, con artists, and liars out of the system.

EXAMPLE 4: BUILDING ECONOMIC INFRASTRUCTURE

With smart contracts, anyone can build entirely new economic infrastructure that enables us to buy, sell, and do business in ways we can only begin to imagine right now.

Smart Contracts and Financial Instruments

Virtually every form of financial instrument that we have now can be implemented as a smart contract. As a result, anyone can look at available contracts, read them, forward them to their lawyers or investment counselors, and decide if they want to invest based on the terms of the contract written in normal legal terms. They can then sign the contract and invest their money. It's all done without any third party such as a broker.

This idea has huge impacts on Wall Street, which has clearly become disconnected with Main Street. The Federal Reserve Bank has used Quantitative Easing to prop up the markets on Wall Street. Billions upon billions of dollars of taxpayer money have been poured into those companies to keep the market at artificially high levels.

But what if most people don't use Wall Street for investing? What if they can do it all on their own? Then suddenly, funneling you hard-earned money to Wall Street doesn't seem like such a good thing for the government to do. Hopefully your taxes can do down as a result.

Other financial instruments that could be implemented through smart contracts include, but are not limited to, the following.

- Private equities
- Public equities
- Bonds
- Derivatives
- Voting rights associated with any of the above
- Commodities

Every type of financial instrument can be added to this list.

Smart Contracts, Transparency, and Anonymity

Smart contracts also provide for full anonymity or full transparency. This is an either/or proposition. You can have full anonymity or full transparency but not both. However, if you want to mix the two, you can. If you want less than full anonymity and less than full transparency, then you can find the balance in between those two that is right for you. It's like a slider with full anonymity at one end and full transparency at the other. You can set the slider anywhere you want to between the two extremes.

Companies and organizations can use smart contracts and side chains (discussed in Chapter 11) to provide full transparency for all of their activities and thereby gain the trust of investors, customers, and the public.

Blockchain IPOs

A company can use digital currencies and smart contracts to provide full transparency into the sale of its stock. You would be able to see how many shares of stock have been issued, how many have been sold, what price they are selling at, and what rate they are selling at. And you would see this 24x7.

In other words, it's possible to build a direct competitor to Wall St. and have companies do their IPOs on digital currency blockchains rather than through traditional means.

In addition, as companies put their transactions, IPOs, and so forth into the blockchain to provide greater transparency, you'll be able to see for yourself exactly how you should invest. The blockchain can provide real, truly perfect data about how a stock is doing in the market at any instant in time because its entire history is available on a second-by-second basis.

The essence of the free market is having accurate information. Putting that information into the blockchain means that everyone has a more level playing field for competing in the marketplace.

Your Personal Transaction History

Another nice feature of blockchains and smart contracts is that you can have anonymity in the marketplace while having all of your transactions available to you transparently. Because your computer can see all of the transactions and investments you've ever done in your entire life, it can analyze your spending habits, help you invest, and help you save. It's easy for software companies to cheaply sell artificially intelligent applications that do all this because they can take advantage of the power of the blockchain.

Because your wallet software tracks all of your transactions, you can literally hit a single button and generate all of the information you need for your annual tax returns. The blockchain essentially provides you with a fully automatic, free accountant.

Of course, most individuals will want their transaction history to be private. But many companies may want to provide a fully transparent transaction history for their investors, customers, and so forth. With the blockchain and smart contracts, they can do this if they choose. And it's all available for free.

EXAMPLE 5: AD HOC RESOURCE CONTRIBUTIONS

Chapter 9 already presented a way that digital cash can be used to sell your unused computing power when your computer is idle. But with smart contracts, you can also sell other resources in an ad hoc manner.

For example, imagine a company that provides cars that you can rent for short trips by simply paying with your digital wallet. Once you have paid, you can start the car and drive where you want to. The company leaves these cars all over town and tracks them with GPS units. They may provide financial incentives for renters to return the cars to specific locations where the company can fill them with gas and check the cars for damage. But the point is that renting a car becomes a trivial effort compared to what happens today. And because you don't have to spend lots of money to rent a car, you can rent one for very short trips instead of taking a cab.

EXAMPLE 6: PREDICTION MARKETS

Most of what we consider to be economic infrastructure is just a fancy form of betting. It's true. We call investments like this "financial instruments" or "prediction markets" but they're really just bets.

For instance, if you buy pork belly futures, you're betting on the price of pork bellies six months or a year from now. Smart contracts are the perfect tool for creating this type of financial instrument. You don't need Wall Street firms to do this at all. A smart contract can monitor the price of pork bellies on the various exchanges. It can manage a single investor or a million investors. It doesn't make any difference.

Prediction markets are a way of "betting" that a particular event will happen. Wall Street uses prediction markets extensively. The blockchain, combined with smart contracts, can make prediction markets so ridiculously easy to build that you can create a prediction market for almost anything. Here's an example.

Internally, the search mega-giant Google uses semi-anonymous prediction markets to determine when it will release software. On software projects, like many other types of production projects, managers say they will have their work done by a particular date. But what they say and what actually happens is routinely different. So Google gives participants money to "bet" on when the software will actually be released. Google management has found that employees who might win some money will often be more accurate about predicting the release dates of its software than its managers are.

This same principle applies across many industries and in many business situations. If people can make money, they will put in the effort needed to accurately predict an event. The event might be who gets elected President of the United States in the next election. Or it might be a foreign policy issue. Or it could be whether oil will be found in a particular location.

Whatever you want predicted can be predicted if you are willing to stake some money on it. Digital cash and smart contracts make this a nearly trivial exercise.

SMART CONTRACTS, BANKS, AND SAVINGS

Government economists love to tell us that consumer spending stimulates the economy. For example, Janet Yellen, the current head of the Federal Reserve Bank said, "Our policy is aimed at holding down long-term interest rates, which supports the recovery by encouraging spending"[129]. Government people believe that if they can just get consumers to spend, it will solve any recession problems.

False.

In order to spend, people must produce first. You have to work at your job or sell goods and services through your business in order to have the money to spend in the first place.

The alternative is to increase your debt. But government economists tend to ignore this. They also ignore the fact that if consumers get laid off during a recession (this happens a lot), they naturally conserve their money rather than spend it. The government belief that we can spend our way to prosperity is ridiculous in the extreme.

SAVINGS AND PROSPERITY

How then do we ensure a robust economy?

Traditionally, the people who make the economy more robust and stable are the savers. When you save your money in a bank, the bank invests the money by lending it out. The money you save goes to finance new businesses, car purchases, home mortgages, educations, and so on. It's not spenders, but savers that have stimulated the economy and made it more robust.

---

[129] Rana Foroohar, "Janet Yellen: The 16 Trillion Dollar Woman," Time Magazine, January 20, 2014.

If you doubt this, simply search the Internet using the phrase "celebrities that went bankrupt" and see the results that you get. These people spent like there was no tomorrow. But that didn't exactly work out for them. They had more expenses than income even though their income was large. In their time, each of them was a veritable stimulus machine, if we adopt the government's view of things. But it did not end well. So it is for the economy as a whole. If we as a nation spend more than we take in, there is only one possible outcome and it is not a more prosperous nation.

Let me repeat, **it is impossible for us to spend our way into prosperity**. Common sense says that if we save we will have a more stable and prosperous financial future.

SAVINGS AND SMART CONTRACTS

But if we put our money into smart contracts rather than banks, doesn't that mean that we're not saving?

Quite the opposite.

The smart contracts offered by blockchain systems enable you to do the same thing with your money that banks do. It just eliminates the middleman and puts you in charge. Just as the bank invested your savings, so you can invest it on your own and get the profit that the bank normally takes. This is great for the economy and great for you.

BANKS AND SMART CONTRACTS

If you don't want the burden of managing your own savings, you can just let the bank do it. Forward-thinking banks won't be put out of business by smart contracts. Clever banks will use smart contracts as a means of simplifying their lending process and driving their costs down.

It's simple for a bank to start a lender's club that all of their customers can participate in. Lenders in the club would deposit their money with the bank. The bank could then put out smart contracts whenever someone comes in for a loan. They would manage the process and reap some rewards for doing so. To you and I, this would look very much like what the bank does now. We would basically be paying them to be our money managers because we don't want to do it ourselves.

BUT NOT JUST BANKS

If you *do* want to manage the investing of your own savings, smart contracts let you do that. Or you can have someone other than the bank do it for you. The most likely scenario for that is a non-bank entity would start a lender's club.

For example, suppose my accountant, I'll call him Bob, offers a new service of matching borrowers to lenders. Note that he's not actually lending or borrowing. He just matches them up by running a lender's club.

Before anyone can borrow or lend, Bob the Accountant has them submit their personal financial status information to him. Bob then vets the lender and makes sure she can really lend what she says she can lend. He also vets borrowers just as a bank would. Borrowers and lenders pay Bob a fee for this when they join the club. It's part of how Bob makes his money.

In the process of joining the club, each borrower and each lender is assigned a reputation score based on their past history. That score can go up or down depending on how the lender or borrower performs in the lending club.

Lenders can then go through the list of borrowers that were vetted by Bob. The lenders examine the contracts that borrowers are offering. They look at each borrower's reputation score. If it's good and they think the contract is a good investment, the lender signs the contract and lends the money. There are probably other lenders that lend money on the same contract. So one lender might lend $1000 and another might lend $100. Lenders will keep lending on the borrower's remaining balance until the requested amount is all borrowed.

On a regular basis, probably monthly, the borrower makes payments directly to the smart contract. The lender gets a part of that payment. The lender's chunk of the payment is directly proportional to the size of the lender's investment. Here's what I mean by that.

Suppose that the borrower put out a $100,000 smart contract for a loan. The smart contract couldn't find a lender for that so it automatically put out 10 subcontracts of $10,000. Some of those were signed by lenders, but the original smart contract couldn't find lenders for the whole $100,000. It therefore automatically divides the remainder into $1,000 contracts. At that rate, it was

> "At its core, bitcoin is a smart currency, designed by very forward-thinking engineers. It eliminates the need for banks, gets rid of credit card fees, currency exchange fees, money transfer fees, and reduces the need for lawyers in transitions... all good things"
> – Peter Diamandis

225

able to get a few more people to lend money to the borrower. But it still couldn't get the whole $100,000. So as you might expect, it divided the remainder into $100 contracts. Let's imagine that our lender lent $100 to the borrower. The lender's $100 is one one-thousandth of the loan amount. So when the borrower makes his monthly payment, the lender in our example gets one one-thousandth of that monthly payment. The smart contract automatically sends the lender his portion of the borrower's monthly payment.

If this is a home loan, the monthly payment might be about $800, depending on a lot of different factors. But let's suppose the payment is $800 per month. So each month, the lender in our example makes 80¢. That doesn't seem like much, but remember that the lender only invested $100. Over the life of the loan, that $100 will become about $250-$300. In other words, the lender has created an income producing asset.

Under our current system, your savings gets invested in income producing assets that are owned by the bank. With smart contracts, you can own your income producing assets directly and make more money than if you used a bank to invest your savings.

The nice thing about this is that any trusted entity can start a lending club. Bob, my imaginary accountant, was able to start his lending club because he knows how to audit the participants. For every contract that gets signed, the contract automatically gives Bob a small fee out of the monthly payments. Usually this fee is not more than a few pennies. In fact, it may even be less than one cent. But for Bob, it all adds up.

Could an insurance company start a club that would enable people to invest in insurance policies for the company's clients? You bet. Investment can happen all over the economy. And that is what *really* makes the economy recover from recessions, not spending.

But wait, what if the borrower doesn't pay? Doesn't that mean you've lost your investment?

Maybe. But there are ways to handle that.

First, if the borrower doesn't pay or is late with his payments, his reputation score goes down. He won't be able to borrow again because everyone will see him as a bad investment. And it's likely that his reputation score will follow him around just the way a credit rating does right now.

Second, the contract can stipulate that the borrower buy an insurance policy on the home that, at the very least, repays your initial investment if the borrower defaults on the loan. The insurance, of course, is handled through another smart contract. This will raise the borrower's monthly payment slightly. But that's what happens now anyway. Part of a borrower's current monthly payment on his home loan goes to provide homeowner's insurance already.

The upshot is that there is some risk in investing. But there is some risk in giving your money to a bank as well. History has shown that people who put their money into a bank's savings account *can* lose it. Even though depositor's money is insured by the Federal Deposit Insurance Corporation (FDIC), that insurance only covers up to $100,000. And if there are a lot of bank failures that happen at once, you will definitely not get your money back.

So no matter what you do, there is no sure thing. You have to decide whether you want to invest your money through your bank or through smart contracts. It's a matter of what reward you want and what risk you're willing to accept.

SMART CONTRACTS DEMOCRATIZE THE ECONOMY

As you can see from the preceding examples, smart contract democratize the economy in ways that nothing else has ever done before. They rewrite the power structures of our civilization and scale them down to the level of the average individual.

Any individual, even those of modest means, can invest like banks do and create income producing assets for themselves. Yes, their investments are smaller than the bank's and yes, their returns are smaller. But in the blockchain economy, more people have the opportunity to invest than they do now. This means that more money flows out to the average person and not to the centralized power structures (banks, Wall Street, the government) like they do now.

The fact that nearly everyone can invest and pretty much everyone can become a stakeholder in the blockchain economy makes the economy far more democratic than it is now. It gives a greater voice to the average person because the average person doesn't need the big corporations and the big government agencies. Virtually all of the functions of Big Business and Big

Government can be replicated through free or inexpensive smart contracts. And those contracts are all voluntary. No one makes you sign. So everyone essentially votes with their signatures or their wallets. The blockchain economy is far more connected to the average person than the dollar economy. In other words, it's an economic democracy.

PATHS OUT OF POVERTY

As you look at smart contracts, you can begin to see the paths out of poverty that they provide. There are ways to accomplish all of the things same tasks that major corporations do by "crowdsoursing" those tasks. That is, hundreds, thousands, or even hundreds of thousands of investors can accomplish the same things that major investment groups do.

Nearly anyone can invest in the blockchain economy. The returns that people get provide them with income over and above what they make at their jobs. Even if their income from their investments is nominal, it adds up over time. If people are smart enough to keep reinvesting most of their profits, they can build a better life for themselves and their children. Smart contracts give even the smallest investor the chance to create something better for themselves.

In addition, smart contracts create new income opportunities that never existed before. For instance, in Example 1, I showed how anyone can provide Internet access to people passing by on the street. Although no one will get rich this way, many people can make some money. So it's worth investing a couple of hundred dollars to set it up. And a lot of people can spend that kind of money. Actually, people in even the poorest neighborhoods sometimes spend more than that on name brand sneakers or a fancy smartphone. So it's not unreasonable to expect that many people with limited funds could supplement their incomes this way.

And of course, providing mobile Internet access is not the only new type of income opportunity that smart contracts create. In a world of smart contracts, we'll see new chances for income all across our economy. People will be able to easily deploy multiple types of resource contributions to various endeavors and earn money that way. They may not make much money at any one of them. But that's ok, because it all adds up. And it adds up especially quickly if people are willing to take their profits and reinvest most of them rather than spending them.

Let's say that an inner city young adult who lives in an apartment right above a busy intersection scrapes together $200 to buy the equipment needed to provide mobile Internet access to everyone passing by. The access point is largely "plug and play" so he really doesn't need to put any effort into running it. It just makes money for him while he goes about his day.

Now suppose that after paying for the electricity to run his equipment, our inner city young adult makes only $25 per month. Not a lot, is it? But let's say he invests $20 a month of that in investments through smart contracts. Imagine that he signed up for an investment group online.

Imagine that over the life of each contract he invests in, his $20 investment gives an average return of $40. This return includes any losses he may suffer from bad investments. If our young adult keeps investing $20 a month for 30 years, he'll invest $7,200. That's not a lot to invest. And all that money comes to him from an income source that doesn't take any effort to run.

> "Money is a collective agreement. If enough people come to the same agreement, what they agree upon becomes secondary, whether it be farm animals, gold, diamonds, paper, or simply a code. History proves all these cases to be true. Who knows what the future is going suggest to us as money, once we see digital currencies as ordinary?" — S.E. Sever

But wait, that's not the whole story. Let's say that his investments return double what he invested after ten years. So $20 becomes $40 after ten years. If our investor is smart, he'll just reinvest his $40 because in ten years, it'll become $80. And ten years later his investment becomes $160. And that's from an original investment of $20. By reinvesting his profits, our investor can make them grow nicely over his lifetime.

Let's take a closer look at how this works. His first year, our investor invests a total of $240. If he keeps reinvesting, after thirty years pass his $240 becomes $1,920. The second year, he invests another $240. He does it again the third year, and so on.

Our investor started investing when he was 25. He keeps reinvesting until he's 55. Then he stops. All of his investments give their final returns 10 years later when he's 65. How much money does he have? The answer is $57,600, which is a very nice nest egg that was gained from just running an Internet access point from his apartment.

But what if he works hard and creates five or ten such income producing opportunities that yield about the same return as the Internet access point? Well, suddenly he has a few hundred thousand dollars to retire on. And that's not bad at all.

The point here is that paths out of poverty are much more possible and more plentiful in the blockchain economy than anywhere else. With work, education, and some reasonable money management skills, people can lift themselves from the kinds of abject poverty we see growing around the world.

It's important to realize that digital currencies aren't a get-rich-quick scheme or a handout machine. The blockchain system is a toolset that allows us to build a better future for ourselves and our posterity.

THE EXPANDING MIDDLE CLASS

Under the blockchain economy, there are some people who will definitely get rich. That's good because they'll get rich by providing new goods and services that no one ever thought of before. In a fair economy, people get the rewards they deserve for innovation and entrepreneurialism.

Neither the blockchain nor digital currencies can keep people from falling into poverty. It will always be the case that some people will become poor. They may become poor because they squandered their money, or because they are lazy, or even because they're genuinely stupid. Or it may be bad luck, poor health, some sort of addiction, or any number of other reasons.

Well welcome to real life. In the real world, this happens. There's *nothing* that can be done to prevent it. And often, failure is what drives people to success. When we cut people off from failure, we also cut them off from success.

But in the blockchain economy, there is nothing like the fractional reserve banking and the government currency monopoly that we currently deal with. Both of these institutions *force* poverty increase over time. There is no other possible outcome. In the blockchain economy, people can fail but their failure is theirs. It's not forced on them by an unjust economic system that enriches the elites at the expense of everyone else.

The result of the blockchain economy, with its myriad of business and investment opportunities, is predictably a thriving middle class. The level of income disparity we see currently will decrease dramatically because the power structures driving money into the hands of the elites will be gone. So there will be fewer super-mega-ultra rich. And by that, I mean the kind of rich people that make Barak Obama (estimated net worth about $42 million), Al

Gore (estimated net worth something more than $120 million), and Mitt Romney (about $250 million) look like modest income earners.

There will still be plenty of opportunity for becoming wealthy. But the vast majority of people will be neither ultra-rich nor ultra-poor. They'll be in the middle class.

And a thriving middle class is exactly what a healthy economy should produce.

THE HUMANE ECONOMY

The invention of the blockchain system gives us the opportunity to create an economy that is not just fairer than what we currently have, it also gives us the chance to create something more humane.

Let's look at an example to see what I mean.

INTEREST IS EVIL

Back in Chapter 2, I showed that when banks charge interest on loans, it concentrates power and money to them at the expense of everyone else. Interest always does this. It never sleeps. It knows no mercy. There is nothing kind about interest.

Does this mean that we should not charge interest?

Yes. With proper innovation, it is possible to completely eschew any type of financial instrument that charges interest. Charging interest always requires people to pay back loans with money that has not been created by the economic system. People trying to pay back interest on a loan do so by competing for an ever-shrinking supply of money that is circulating through the system. This is one of the primary causes of poverty in our world today.

Of course, modern banking is built almost entirely on interest-bearing financial instruments. They are so common that they seem natural to us. In fact, even most people in the financial industry would be hard pressed to figure out new types of financial instruments that do not depend on charging interest. But the reality is, it's all been done before.

Prior to the 1600's, most cultures equated charging interest with usury, which was strictly forbidden in the Bible, the Koran, and other sacred writings. During the 1500's and 1600's, people redefined usury to mean charging "excessive" interest. The laws of most Western nations were built on this idea.

To build a better future, we can look to the past to see what kind of financial instruments were used during times that prohibited interest.

EXAMPLE: THE UNMORTGAGE

A good example of an innovative financial instrument that does not require the payment of interest is what I call the "unmortgage." This financial instrument performs the same function as a mortgage but doesn't involve interest. I call it the unmortgage because there is no English word for it[130].

Before we dive into this, I want to say that it is completely possible to implement the unmortgate with US dollars or any other kind of currency. However, banks will laugh you out the door if you try to present them with the idea I'm explaining here.

Because mainstream financial institutions are so heavily regulated and so resistant to change, the unmortgate and other non-interest-bearing financial instruments are more likely to find widespread use in the blockchain system. It's a natural fit given that people in the blockchain system will need innovative ways to invest their money.

IT STARTS OUT NEARLY THE SAME AS A MORTGAGE

Suppose you want to buy a home. The first step in the unmortgage process is to find a local investment group a that handles unmortgages. They approve you just as if they were a bank.

In this example, we'll say that the home costs $200,000 and you are required to put down 10% on the home.

So far, this sounds like a regular mortgage, doesn't it? However, here's where things start to get different.

---

[130] In the Muslim world, this type of mortgage would be called a halal mortgage.

THE PARTNERSHIP BUYS THE HOUSE

The investment group buys the house with credits that members have deposited in its investment account and with your $20,000. The investment group is 90% owner of the house and you own 10%. You have entered into a business partnership with the investment group. You are not debtor and creditor, which is of necessity an adversarial relationship.

YOU RENT THE HOUSE

You now rent the house from partnership and move in. The monthly rent is approximately the same as the payments you would pay with a mortgage. When you make your monthly payments, 90% of the rent goes to the investment group. The investment group keeps some of that for the service of running the group. The investors get the rest.

The other 10% of the rent you pay is yours. However, you don't put it back in your pocket. Instead, you pay your 10% to the investment group for a greater share of ownership in the house. So every time you make a rent payment, the investment group gets 100% of the rent. That is how it makes its money without charging interest.

Each time you pay rent, your 10% that you pay to the investment group increases share of the ownership until you are the full owner of the house. At that point, the partnership terminates. You are the homeowner. You have paid about the same amount of money that you would under a mortgage. The investment group has made the same amount of money that a bank would make with a regular mortgage.

So how is this different than a mortgage?

THE INVESTMENT GROUP MAKES A PROFIT THROUGH RENT RATHER THAN INTEREST

A 30-year unmortgage will make about the same amount of money for an investment group as a normal 30-year mortgage. Because they are getting the same amount in rent that they would get by lending you the money with interest, the profits for the investment group are about the same as if they lent you the money.

But the unmortgage does not require the payment of interest. It does not have the same debilitating effect on the homeowner that a mortgage does. In addition, it provides more humane alternatives than a mortgage does should something go wrong in your life.

WHAT IF SOMETHING GOES WRONG?

To this point, the money involved in an unmortgage is about the same as the money involved in a mortgage. This seems like an awful lot of trouble to go to just to avoid interest–especially if it doesn't save you any money. So why would you want to bother with an unmortgage?

Suppose something bad happens to you. You become unable to make your payments. If you have a mortgage, everything is stacked in favor of the bank, so you lose *everything*. You lose all of your equity. You get evicted and you lose your home. You loose your credit rating. You lose your ability to buy another home for years.

But that's not the case if you have an unmortgage. If you can't pay your rent every month, the investment group must evict you for nonpayment. However, you do not lose your equity in the house as you would with a normal mortgage. The investment group does not repossess your house. They can't–you are a part owner.

Instead, the investment group just rents the house out to someone else who can pay the rent. The partnership continues. You still own your share of the house. They still own theirs. Each month, the renter is paying rent. Part of that rent belongs to the investment group and part belongs to you. You are still paying your part of that rent to the investment group so your share of ownership in the house is still growing. Unlike a regular mortgage, where you must necessarily lose everything, you are still on the path to home ownership.

Meanwhile, you do your best to recover from your financial setback. Even if you can't move back in for many years, you are still on your way to being a homeowner for as long as the investment group can keep the home rented.

The unmortgage is a distinctly humane approach to home ownership by comparison to what we have now. There is a very real possibility that you will be able to get back on your feet and start making rent payments again. In that case, the investment group can let you move back in when the lease with the current tenant expires. You resume renting and all continues as before.

PARTNERSHIPS INSTEAD OF LOANS

With the unmortgage as an example, it's easier to see how we might create an economy that is not based on interest. For example, it might be reasonable for the investment group to enter into partnerships with companies that require fleets of cars. The company can start out by putting 10% down on its fleet and then rent the fleet from the investment group. Eventually, the company will own the cars.

Other types of partnerships might also be possible. For instance, do factories really need to own their machinery? Might it be profitable for investment groups to create a business that owns such machinery and leases it out to factories or has unmortgage-style partnerships with them?

In any case, avoiding interest-based financial instruments avoids the crippling weaknesses of the fractional reserve banking system. Investments based on joint ventures, equity shares, claims against future production, and revenue sharing can be as profitable as interest-bearing financial instruments and be healthier for everyone involved in the long term. You are really only limited here by your imagination.

THAT'S NOT ALL FOLKS

So you've seen how smart contracts can fundamentally change our world and enable us to reinvent the economy in fairer and more humane ways.

But there's more.

Really, there's more. The next feature of the blockchain may turn out to be just as impactful as smart contracts. So let's forge on to Chapter 11 and see what that feature is.

11 SIDE CHAINS

Blockchains can also be made to work with other blockchains, which are called *side chains*. A side chain is a blockchain that is based on and works with the main blockchain. So you can create a main blockchain that mints your currency and have an unlimited number of side chains that work together with the main blockchain to add new features and services.

Side chains are great places for innovators to do whatever they want without interfering with the main blockchain. Currencies and tokens created on side chains are directly exchangeable for the coins on the main blockchain. In fact, if the main blockchain has a built-in distributed currency exchange, you can buy and sell tokens from side chains them in the main blockchain's distributed exchange.

Having side chains enables innovators to do anything they want to do without the possibility of disrupting the main blockchain. If someone's experiment fails, the main blockchain is not cluttered with data from the experiment. It's all on a side chain that is probably ignored and unused. Eventually, it will disappear from the system entirely.

If side chains are used to create currencies, their values can be pegged to the value of the coins from the main blockchain. A currency in a side chain can use the same method that the main blockchain uses to regulate its coin supply and value. Or its creators can choose an entirely different method. This enables anyone to use side chains as an easy way for creating innovative new currencies.

In fact, it is hoped that side chains will become the basis for industry-specific currencies that will optimize the flow of money through particular industries and create new investment opportunities. Combining both a side chain and smart contracts could lead to not only industry-specific currencies, but entire economic infrastructures based on those currencies which would enable new innovations, new investments, and new opportunities for profit that we do not now have.

TYPES OF SIDE CHAINS

In the world of blockchains, there are two types of side chains: one that is pegged to the main blockchain and one that is interoperable with the main blockchain.

PEGGED SIDE CHAINS

Side chains that are pegged to the main blockchain produce coins or tokens that are actually backed by the coins from the main blockchain. Typically, each coin in a side chain contains a coin (or a portion of a coin) from the main blockchain inside of it.

A pegged side chain mimics the traditional gold standard style of producing money. For example, the US dollar used to be backed by gold. In that same way, pegged side chains are backed by coins from the main blockchain so coins from the side chain have instant value.

It Costs You a Bit

The process of pegging side chain coins to coins from the main blockchain actually transfers coins from the main blockchain into the side chain. The pegged coins from the main chain can't be used for anything other than providing value to the side chain coin. That is, pegged coins from the main chain can't be spent independently of the side chain coin.

Pegged coins generally cannot be transferred back to the main blockchain if the side chain coin goes out of existence[131]. There are significant technical difficulties with passing the pegged coins back into the main blockchain. Therefore, most blockchain systems simply don't allow it.

> "Today, we google for everything, mostly information or products. Tomorrow, we will perform the equivalent of "googling" to verify records, identities, authenticity, rights, work done, titles, contracts, and other valuable asset-related processes. There will be digital ownership certificates for everything." - William Mougayar

But if what you are looking for is a way to create a currency with instant value, then a pegged side chain can be just what you need, in spite of the cost.

---

[131] There is work being done to try and develop methods of returning the pegged coins back to the main blockchain. However, transferring coins back to the main blockchain is highly problematic. It is difficult to make it work properly. In fact, it is so difficult that it is unlikely we will see it in blockchain systems any time soon.

They are Money Too

Pegged side chains are designed specifically to be a type of currency. For example, an industry-specific currency would use a pegged side chain to provide its coins with value. It would then add industry-specific usage and transaction rules that are not shared by the main blockchain.

For instance, you could use a pegged side chain to finance a distributed, decentralized storage system in which everyone who wants to can make money by providing disk space. This would essentially bring the much-ballyhooed "cloud" down to the street level. That is, everyone can participate in a secure, cloud-based storage system that is completely decentralized and make money by doing so.

The pegged side chain would enable the system to use the side chain's coins for specific purposes within the decentralized, cloud-based storage system. And the coins could only be used according to a specific set of rules that were not applicable to coins from the main chain. But the side chain's coins would have instant value because they contain a coin from the main chain inside themselves.

INTEROPERABLE SIDE CHAINS

Interoperable side chains are blockchains whose coins are not backed by coins from the main blockchain. They can be used as tokens in private trading systems, they can store data, they can be discount coupons, or they can be used in almost any other way that you can think of.

Unlike pegged side chains, interoperable side chains are nearly independent of the main blockchain. Interoperable side chains can have nearly any rules that you want them to have as far as what transactions are allowed, how transactions are performed, and so forth.

The most useful feature of interoperable side chains is that their coins are formatted so that they can work with many of the features of the main blockchain. For example, coins created on an interoperable side chain can be bought and sold on the main blockchain's decentralized coin exchange. They can also be used in smart contracts, which is another feature of the main blockchain.

> "Cryptology represents the future of privacy [and] by implication [it] also represents the future of money, and the future of banking and finance." — Orlin Grabbe

Using Interoperable Side Chains

How would you use an interoperable side chain? Suppose you want to create a public system that anyone can use to sell downloadable digital content such as music, books, pictures, software, and so on. There are a number of ways you could approach this. But one way is to use an interoperable side chain in which the digital content is "stamped" into side chain coins.

Let's call our side chain for this example MediaCoin. Miners would mine MediaCoins and sell them to authors, artists, and other content producers. Content producers would them perform a transaction that permanently stores their content into the MediaCoin that they bought in an encoded form. The content producer could then use a smart contract to provide access to the coin. The smart contract would issue five tokens called AuthorizationCoins, or AuthCoins for short.

Content producers would sell their content from their web sites. Let's say I'm using your MediaCoin system to sell books. When I sell a book online, my customer is actually buying a MediaCoin containing my book, plus a smart contract. The contract enables my customer to authorize up to five devices for the content. That is, my customer can read my book on up to five devices. The contract would allow my customer to deauthorize old devices and authorize new devices, but there could never be more than five authorized devices at any one time. The contract does this by putting an AuthCoin on each authorized device. It will never allow more than five devices to contain an AuthCoin.

Your MediaCoin system would enable anyone to sell content publically on the MediaCoin blockchain. Anyone could search the MediaCoin blockchain for books, music, or other content they were interested in. The MediaCoin blockchain would encourage people to provide resources by rewarding them with blank MediaCoins, which they would then sell to content producers.

If it was more efficient, you could also implement AuthCoins as an interoperable side chain. In that case, they would not be issued by the smart contract. But the smart contract would buy them when the contract created and it would manage them for as long as the customer owned the content.

If you think about it, the impact of this is huge. For example, if you want to make and sell a TV show, you can simply do it without going to any TV networks.

If you want to make your living as a news reporter for your area, you can. Simply start publishing your own news stories. If they are well done and timely, you will gain a following and bring in a good income. If not, well perhaps you can think about becoming a pastry chef.

From the buyer's perspective, a blockchain for selling digital content is wonderful. If you want to watch TV shows, you can simply search[132] for them in the blockchain, buy access for a small fee, and watch the show. Likewise, you don't need news services any more. You can simply search the blockchain for news stories and buy access.

Advantages of Interoperable Side Chains

There are many advantages to a MediaCoin-style system. First, you don't have to put up with the onerous policies of eBook and digital music publishing companies that you must deal with today. These companies often gouge into your profits in exactly the way that traditional publishers have.

Also, you get an automated sales and content management system at no charge to you. You just have to connect to it from your web site.

In addition, anyone could create their own custom book or music store with this system. For example, if I wanted to create an online bookstore that specialized in books for teens and young adults, I can simply put up a web site and start searching the MediaCoin system for good books.

I would then contact the individual content producers and provide them with proof that I am a business that resells books. Instead of the standard retail smart contract they use when they sell to customers, they would use a wholesale smart contract that enabled me to buy their books at a discount and resell them to customers. This way of using interoperable side chains would provide an entire ecosystem of independent content producers and resellers with the means of establishing a flourishing digital content market.

STORING DATA IN A SIDE CHAIN

Another good use of a side chain is for data storage. For example, you can create a side chain that stores documents. Using a side chain in this way

---

[132] It's likely that when a blockchain for selling digital content comes online, you will see specialized programs become available that search the blockchain for particular types of content. They may even make recommendations for you based on what you've bought before.

means that the documents are time stamped so that anyone can prove that a document contained specific content at a specific time.

SIDE CHAINS AND GOVERNMENT TRANSPARENCY

A side chain of provably time-stamped documents would provide a decentralized repository for public-facing documents. For instance, such a system would be an excellent way of providing transparency in government.

"Today thoughtful people everywhere are trying to understand the implications of a protocol that enables mere mortals to manufacture trust through clever code. This has never happened before—trusted transactions directly between two or more parties, authenticated by mass collaboration and powered by collective self-interests, rather than by large corporations motivated by profit.

It may not be the Almighty, but a trustworthy global platform for our transactions is something very big." - Don Tapscott and Alex Tapscott

By requiring the government to store all but the most sensitive documents in such a blockchain, we would have a total and complete accounting of everything the government is doing and how it spends public taxpayer money. Imagine how different things would have been in the 2008-2009 bailouts if all documents relating to the bailouts had been stored in a publically available blockchain where anyone could do a full audit of how the funds were spent?

Other examples of this might be land deeds and treaties. There have been instances in history where deeds or treaties were altered by corrupt government officials to seize the land owned by racial and religious minorities[133].

Forcing the government to put all of its treaties, deeds, court records, and so forth into a public blockchain would mean that they could no longer forge, destroy, or alter important documents[134].

In fact, blockchains are perfect for keeping the government honest. As bitcoin advocate Paul Snow has famously and humorously noted, "Public

---

[133] This was not unknown in the American South after the Civil War. Mobs sometimes chased black people off their lands and then just wrote up new deeds. In the early 1800s, the Mormons experienced similar abuse by local governments. Native Americans, who often couldn't read English when treaties were written, were also the targets of easily-altered treaties.

[134] Altering government documents happens today. A lawyer has recently accused the SEC of destroying documents related to convicted felon Bernard L. Madoff, as well as Wall St. firms that were culpable in the 2008-2009 financial meltdown. The US State department has recently accused Hillary Clinton's staffers of destroying documents related to the Benghazi massacre that resulted in the death of a US ambassador. Lest we be accused of being partisan against the Democrats, let us recall that Oliver North, who worked for the Reagan administration, admitted in sworn testimony that he destroyed documents related to the Iran-Contra scandal of the 1980s.

ledgers (i.e. blockchains) tend to be honest because it's very hard to know today what lie I want to tell tomorrow."

SIDE CHAINS AND IDENTIFICATION

Blockchains can provide a simple, inexpensive, and unforgeable way to create identification papers. This can eliminate the need for government IDs, passports, and so forth.

Using information that only you can generate, you can use a blockchain to encrypt information on an ID card. The information, which can include biometric data such as your picture and fingerprint, is encrypted using a hash function and stored as nothing more than a number. That number can then be put into a bar code or QC code.

Because only you can generate the resulting hash value, you have a way to prove who you are no matter where you go. So your ID can potentially be universal across all national boundaries.

And like other side chain applications, blockchain IDs are self-supporting. No tax money is required to create them. The system pays the people who contribute resources to it.

SIDE CHAINS AND COMMON PUBLIC RECORDS

Side chains are the perfect place for all common public records.

Business Licenses, Incorporation Records, and Dissolution Records

Currently, you must have a business license from the government to engage in independent economic activity of any kind. Specifically, you must have a business license to do business.

Does that make sense to anybody? Why in the world should I have to get permission from the government to engage in economic activity? Never before in the history of the world, except in totalitarian governments of the past, have people been required to have government permission for buying, selling, producing products, and offering services. It makes no sense that you or I should have to have permission from the government to create our own businesses.

Ostensibly, the government justifies this requirement by stating that it needs to keep track of businesses in order to collect taxes. The government thinks that it and it alone can issue you an identification number that you must use to pay your taxes. Well we wouldn't want to stand between a politician and the taxes he thinks he deserves.

Nevertheless, this is no justification for surrendering control of our economic freedom to the government. It is a straightforward endeavor to create a side chain that is used for registering businesses. Everyone can supply the required documents, store them in the blockchain, and obtain a unique ID number that they can then use to pay taxes with. The government need have no say in it at all. We should all be able to form whatever businesses we want without government approval. It's called being free.

Likewise, there is no reason to file business incorporation or dissolution records with the government. If it needs to access such records, it can do so through the publically available side chain.

Vehicle Registrations

As with business licenses, there really is no reason for the government to be running a registration database of all of the vehicles we own. They are our personal property and we should not have to pay fees to own and use them.

A vehicle registration system, which is easy to implement in a side chain, can be privately implemented, privately run, and be completely self-supporting. Registering your vehicle in a side chain would prove your ownership of the vehicle at all times. Of course, such a system is especially useful if your vehicle is stolen.

Many More

Other records that can be turned over to blockchains include, but are not limited to, the following.

- Birth certificates
- Death certificates
- Health/safety inspections
- Court records
- Building permits
- Regulatory records

SIDE CHAINS, COPYRIGHTS, NOTARIZATION, AND PATENTS

Another use of a time-stamped side chain could be the registration of copyrights for content such as videos, movies, books, articles, music, and so forth. Right now, you must register your copyright in a time-consuming process with a government agency. However, there is no reason for taxpayers to fund such a system. A blockchain-based system that does the same thing would be self-supporting and require no taxpayer money.

The first person, company, or organization that puts the content into the blockchain gets the copyright. And because that entity (person, company, or organization) can prove that they were the first one to post the content into the blockchain, they can use the blockchain as evidence in court whenever anyone violates their copyrights.

This same approach works for any document that needs to be notarized. To notarize a document, you simply sign it, encrypt it, and store it into a blockchain. Then it becomes part of the immutable blockchain record.

Patent applications and patent grants are another perfect application for blockchain storage systems. They provide a way for anyone to prove that they filed a patent application on a specific day at a specific time and that record can never be altered. And again, it doesn't have to be supported by taxpayer funds. The system pays people for providing resources for the system. It's completely self-supporting.

Note

Storing digital content in a blockchain can make the blockchain quickly become huge. Although this is not that much of a problem given the plummeting prices of hard drives, it is possible that in some instances it can become an issue. Fortunately, there is an alternative. Rather than store the original content, you can store a signed and hashed version of the content into the blockchain. The resulting hash codes take up very little space. But you then have a way to prove that your copy of the content is the original one because only your original signed copy will generate the hash code that's stored in the blockchain.

SIDE CHAINS AND HISTORY

George Orwell superbly said, "Who controls the past controls the future. Who controls the present controls the past."

This statement is based on the fact that governments and other groups have always altered their histories to suit their own purposes. And they chiefly alter the record of the past to help keep themselves in power.

However, documents stored in side chains are immutable and time stamped. And if they are stored in an unencrypted format, they are publically available worldwide.

A Shared History For Humanity

What this means is that for the first time in human history, we have an ordered public record that cannot be altered. We can store historical documents in it and they will always be available to everyone in future generations. No government, political party, or anyone else can ever change the public record of what happened in the past. They can add new documents. And the new documents can even be updated versions of the old documents. But they cannot change the old documents. Nor can they change the order that the documents were added to the blockchain in.

For example, all of the documents associated with a presidential administration can be stored in a historical blockchain. Then all of this information is available to anyone forever. No one can go back and try to alter the record of what happened during a particular administration. Everyone can see the original documents and decide for themselves what they think about the events of the past.

And Even for Games

Although the idea of an unalterable shared history has profound implications for human history, it can also be applied to more trivial situations for profit-based applications. For instance, suppose that I participate in an online game that uses leaderboards. The leaderboards keep my game history and accomplishments. Many online game services use leaderboards in order to give their players "bragging rights" and opportunities to show off.

One way to implement leaderboards is to create a side chain that stores leaderboard information. The side chain would reward players somehow for storing the blockchain on their computers. But because the side chain contains a complete history of everyone's game accomplishments, the game company has an easy, low-cost way of implementing leaderboards on its service.

So whether we're using publically shared histories for nobler purposes like providing humanity with an accurate memory of its history or for profitable purposes like leaderboards in games, we can easily create them using side chains.

SIDE CHAINS AND COMPANY TRANSPARENCY

It was previously mentioned that companies can provide transparency for investors using smart contracts and the blockchain. In fact, the tools that they would be using are smart contacts and a side chain. It is really possible for every company to have its own side chain for storing its entire transaction history, stocks, and so forth. They don't have to, but they could.

Right now, if you want to know the status of a large company, you hire a small army of highly trained accountants to go through the company's finances and perform an audit. In the end, they can only give you an opinion of whether or not the company's finances are on the up and up.

If a company's entire transaction history is available in its own blockchain, or in a blockchain that is dedicated to keeping company transaction histories, then it's easy to perform an audit with the proper software and have a full and complete record of the company's activities.

Of course, every company has information that it wants to keep private. There are ways to store your transaction history in a blockchain and still keep private things private, or at least limited to those who should have access to the information. But the record of the transaction is there and must be accounted for in an audit. Therefore, it's difficult if not impossible to hide damaging information when it must be disclosed, as in a court case.

SIDE CHAINS AND DATA PROTECTION

One of the nice features of documents stored in side chains is that there are many copies of them. Because of the distributed nature of blockchains, any data stored in a blockchain is automatically replicated many times. The blockchain keeps all of the copies of the documents in sync. This is a definite advantage over centralized systems. Large document collections in centralized systems are vulnerable to a distributed denial of service attack (DDoS). Blockchain-based storage systems do not have this problem[135].

In addition, having many copies of important documents spread around the country or around the world means that the documents are safe in the event of a local natural disaster or some other local emergency.

Documents stored in side chains can become the basis of many other innovations that can reduce the size of government, increase the resiliency of our democracy, and provide the basis for innovations that we cannot now imagine.

SIDE CHAINS AND PRIVATE DATA STORAGE

Besides storing public-facing documents, innovators can create side chains to store private information. This information would be "shredded" and then encrypted at least once[136].

How do you "shred" information? Here's a simplified example. But please realize that real "shredding" would be more complex and more secure than this. It would also be distinctly more secure.

Let's suppose that the data you want to store in this distributed, side chain-based storage system is actually your digital cash from a particular blockchain, and that blockchain is the main chain. Now let's imagine that you have four coins from the main blockchain. Figure 11.1 demonstrates the simplest and most basic way of shredding your coins.

	1	2	3	4	5	6	7	8
Qbit 1	1	0	0	1	1	1	0	1
Qbit 2	0	0	1	1	0	0	0	0
Qbit 3	1	1	1	1	0	1	1	1
Qbit 4	0	1	0	1	0	1	0	1

Figure 11.1 Shredding Private Data

---

[135] DDoS attacks on blockchains are possible but very unlikely and very preventable.
[136] It is possible to use multiple layers of encryption to increase the security of a distributed data storage system.

For the sake of this example, we've assumed that there are only eight bits of information in a coin. That's not really true, but it doesn't matter right now.

Notice in Figure 11.1 that each coin is in a row by itself. Each bit of the individual coins is set into a specific column.

In a side chain-based private data storage system, the software would take bit 1 of coin 1 and bundle it together with bit 1 of coin 2, coin 3, and coin 4. So it would grab all of the bits in the first column and put them together. It would then store those four bits somewhere in the system.

Next, the software would take the four bits in column 2 and store them somewhere else in the system. These two pieces of data may be stored on computers on opposite sides of the world.

The software continues by storing the bits in columns 3-8 in exactly the same way.

Storage space in this system would be provided by private individuals and companies. They would be paid for the amount of their storage that the system uses.

Let's imagine that the guy who provides the storage space for the bits in column 1 is able to crack the encryption and decode the four bits. He now knows that they are 1010. What does this get him?

Absolutely nothing.

The four bits are from four different coin. He doesn't know where the other bits for those coin are. He has no way of finding out.

Where are those bits?

Only you can say. If your wallet is integrated with a side chain that provides this kind of private data storage, then it automatically keeps an index of where your money is stored. It isn't possible for anyone else to find out where all the bits for your money are. Only you can reconstruct your money[137].

Even if a hacker is successful in decoding a few bits of your data, he has no way of finding the rest of it. Therefore, he can't reconstruct the original data.

---

[137] In a real system that is built along these lines, your wallet would also have safeguards built in to protect your index from hackers and other forms of theft.

Documents, money, and other information stored in this way would be completely private. There would be multiple copies of your data across the system in the event of a local disaster so that your data can be reconstructed no matter what. And the system is totally scalable. As more people that it, more people and companies will provide services for it because they get paid to offer storage space for data.

SIDE CHAINS, SMART CONTRACTS, AND DECENTRALIZED AUTONOMOUS ORGANIZATIONS

A side chain that stores documents, together with smart contracts, can provide the basis for decentralized autonomous organizations (DAO). DAOs are organizations that people just join, use, and contribute resources to because it's in their best interest to do so. There is little or no central control.

One possible DAO that is enabled by a side chain is a bank. A DAO bank would have no corporate headquarters. It would function according to a set of rules described in documents stored in the side chain. Anyone who participates in the DAO bank can be sued by other participants if they do not participate in the way that is described in the founding documents. In this way, the organization can police itself even if there is no central control. It is even possible that the blockchain and smart contracts could provide the basis for the bank's internal arbitration system, if it has one.

The DAO bank would use automatically pay people for providing services to the system. If too many or too few people try to provide services, the laws of supply and demand would automatically decrease or increase the number of service providers as needed.

Another way that DAOs could be used is to provide roads. A common objection to limited government is, "Who would build the roads?" In considering this question, let's ignore the fact that roads were built and maintained for literally hundreds of years in North America before governments were involved with them.

When a builder decides to put up a housing subdivision[138], it can build the roads. The builder can then use a side chain and smart contracts to create a DAO. The members of the DAO would be the homeowners in the subdivision. The DAO could automatically collect and manage funds, as well as handle other tasks needed. The government need not be involved at all.

SIDE CHAINS AND CODE

For programmers, side chains can be a bonanza.

Why?

Because you can store source code, script code, or even executable code in a side chain and make it accessible from anywhere. And if you're worried about how the code will run on people's systems, you can also embed the compiler in the blockchain. Or you can use an interpreter.

Now combine your ability to make source code, scripts, and binary executables with the ability to store smart contracts and digital content. What can you build? Remember, this will be available worldwide.

Could you make an artificially intelligent blockchain that will bring the financial analysis abilities of Wall Street to the masses? Or could you create a side chain that stores and analyzes the research data needed for advances in medicine and automatically lets subscribers have access? Could you build a side chain that extends most or all of the advantages of blockchain-based currencies to traditional currencies like dollars or euros?

It's all up to you. The possibilities are really endless.

---

[138] Even today, most roads are not built by the government. When a developer puts up a housing division, it builds the roads, as well as all of the utilities. All the government does is make sure they are built to existing standards. For the most part, it is private industry that builds the roads in this country, not the government.

12 THE BLOCKCHAIN AND THE SOCIAL REVOLUTION

With the advent of the blockchain, we are creating a world where anyone can invent a new type of money. And the money we invent can be designed to solve specific social problems. This type of currency is called a *social-purpose currency* or a *complementary currency*.

Social-purpose currencies can solve social problems without raising taxes, redistributing wealth, increasing odious government regulations, or going into debt. Although a statement like this is astounding to most people, it's perfectly true. What's more, it's proven. Social-purpose currencies have actually been created and used in the past to solve otherwise intractable problems.

In this chapter, we'll dive into the realm of social-purpose currencies and get a glimpse of what we can implement with the blockchain. But as you read this, please remember that I'm only scratching the surface of what can be done. If you want to really solve massive social problems, then the blockchain gives you the tools to experiment and find the right path forward. With some innovation, you really can make a difference.

MONEY IS NOT VALUE NEUTRAL

If you dive into economic theories, you'll find that nearly all of them assume that money is *value neutral*.

Value neutral? What does that mean?

This is actually a simple idea. It means that the features of money do not affect how people use it, save it, and spend it.

This concept is absolutely false. The reality is that the features a currency has very profoundly impact how people use that currency.

For example, businesses often implement customer loyalty programs such as airline miles, cashback bonuses, discount coupons, and so on. All of these are

253

just specialized currencies. Businesses have found that the features of a business's customer loyalty currency very directly affect the currency's success or failure.

And if we really look hard at our own money, we see the same thing; it is definitely *not* value neutral.

THE FEATURES OF OUR MONEY

Our money has three primary features. Our money is:

1. A unit of accounting.
2. A store of value.
3. A medium for exchange.

Two of the three features our money has are at odds with each other. In particular, if a currency is a store of value, that encourages people to hold onto it and to never spend it. But if it's a medium of exchange, that encourages people to spend it and not save it. We have schizophrenic money.

Most people think that *all* money has these three features. That is false. Most people also think that all money *must* have these three features. That is false as well.

In fact, there have been multiple currencies used for literally thousands of years that did not have all three of these features.

I am not suggesting that we should change the features of our existing money. That's not the point. The point here is that there have historically been other currencies that did not have these three features. And as a result, people in those times used money differently than we do. Their money was not value neutral just as ours isn't. The features any currency has can and does influence how people use it.

MONEY IN ANCIENT EGYPT

One example of a nation that had money with different features is ancient Egypt. The currency of ancient Egypt was not a store of value. Specifically, it was subject to *demurrage*. Demurrage means that when the current Pharaoh died, the money with his image on it was worth less. The new Pharaoh would mint new coins with his image on them and everyone could exchange the old ones for the new money, typically at a rate of 4 old coins for 3 new ones.

Because you never knew when the Pharaoh might get sick and die, or die in battle, or get killed by an animal in a hunting accident (the Pharaohs were avid hunters), people did not save money. Instead, they spent it rather quickly on buying, building, and improving income-producing assets such as land, livestock, tools, small factories, waterwheels, and do forth. The general population ended up buying or building as many income producing assets as they could. Even if the money they were holding got devalued because the Pharaoh died, they still had new streams of money coming in. As a result, they were very prosperous for literally thousands of years.

Why did the Egyptians stop using their form of money? They were conquered by the Roman Empire, which invented our form of money to enrich the aristocracy. The Romans forced our kind of money on them and their civilization completely changed (not for the better) as a result.

Again, the point here is not to advocate any particular kind of money. It's just to indicate that the features of money directly impact how it is used. That is the case with the money we have now.

SPECIALTY CURRENCIES CAN SOLVE SOCIAL PROBLEMS

Because money is not value neutral, it is possible to invent currencies to solve otherwise intractable problems. This has actually been done in modern times. Here are some examples from Brazil.

THE BRAZILIAN SABRE

Economist and currency expert Bernard Lietaer (one of the creators of the Euro), designed a currency called the Sabre[139] for the Brazilian educational system to help it increase the effectiveness of the money the government spent on schools. At the time he created the Sabre, the Brazilian government was looking for ways to do more for its children without adding pressure to its already strained economy.

---

[139] For more information on Bernard Lietaer and his work on social-purpose currencies, please see Chapter 15 of New Money for a New World, Bernard Lietaer and Stephen Belgin, Qiterra Press, 2011.

The program that Lietaer came up with was surprisingly simple.

1. Create a complementary currency (a social-purpose currency) called the Sabre that is backed by a grant from a national educational fund.
2. Give the Sabre to 3rd graders.
3. The 3rd graders use the Sabre to pay 4th graders to tutor them. A teacher records the transaction.
4. The 4th graders use the currency they've received to get tutoring from 5th graders. The 5th graders
5. This continues on up through the grades until the complementary currency ends up in the hands of the high school seniors.
6. The seniors can use the currency to pay college tuition to participating colleges. The colleges can turn the currency in to the government to get money from the original grant.

In general, the colleges that participated in this program were running at less than their full capacity of students. The program requires them to charge Saber students 50% tuition. Colleges participated because receiving 50% tuition was better than receiving 0% tuition. Once a student was into the college, the colleges were required to continue to accept Sabers for that student's tuition until the student graduated and to not raise that student's tuition.

There are a lot of reasons why this innovative program is so effective.

Magnified Value

First, the Sabre propagates the value of the actual money through the entire educational system without actually spending the money itself. Until the children graduate from high school, the money remains in the bank. But in the meantime, the value of that money is being used over and over again through each grade level until students actually apply it to their college tuition.

Low Overhead

Second, it adds very little overhead on the part of the school to derive the value of the program. Essentially, it's one teacher or staff member sitting in a room with the students while the students give or receive tutoring. That one adult keeps order and records the transactions on a computer when the tutoring is complete. The overhead is slight relative to the massive value being received.

Redesigning Money Equals a Redesigning Education

Third, the Saber educates children in ways that are far more effective than traditional means. Here's why.

Our modern school system was invented by the Prussian military and used to educate boys (because back then it was uncommon to educate girls) in a way that would produce good soldiers. Specifically, it was designed to increase what we now call left-brained activity in students. It depended heavily on rote memorization and absolute obedience.

Education for Marching Soldiers

At the time, soldiers marched in formation across open fields and shot at each other, stabbed each other, etc. These formations were an extremely effective method of fighting given the weapons of the time.

But moving in formations required soldiers to stay in complete and absolute lock step with each other. Innovation, creativity, or questioning orders would result in the whole formation falling apart. That in turn usually ended with the deaths of all involved. So the school system the Prussians designed was specifically formulated to crush innovation, creativity, and free thinking out of students.

> "It is, in fact, nothing short of a miracle that the modern methods of instruction have not yet entirely strangled the holy curiosity of inquiry; for this delicate little plant, aside from stimulation, stands mainly in need of freedom; without this it goes to wrack and ruin without fail. It is a very grave mistake to think that the enjoyment of seeing and searching can be promoted by means of coercion and a sense of duty." - Albert Einstein

The Modern World is Right-Brained

As a result of modern entertainments, most especially video games, students' brains have been undergoing a fundamental shift over the last few decades. We're seeing a profound shift from left-brained thinkers to right-brained thinkers[140]. This is neither good nor bad. It just is.

The change from left-brained thinking to right-brained thinking is most extreme in boys (boys tend to play video games more than girls), and it has corresponded directly with a massive drop in educational success for boys. Our left-brained educational system can no longer effectively teach our right-brained children. This is one reason we have more gifted underachievers than gifted overachievers in our schools today.

---

[140] For detailed information on this change in our children, see Daniel H. Pink, A Whole New Mind: Why Right-Brainers Will Rule the Future, Riverhead Books, 2006.

Teaching Right-Brained Children

For right-brained children, one-on-one teaching is the best way to learn. The worst way to learn something is to receive a lecture about it. In a lecture situation, only about 5% of the material presented will be remembered. Things get slightly better with material that you read. You'll remember about 10% of that. If there's a demonstration of the concepts taught, recall will go up to about 30%. Both discussion groups and practice by doing also increase retention. But the best way to learn something is to teach it. In that case, retention goes up to 90%.

What makes the Saber and programs like it so appropriate for modern children is that it first makes their learning process easier by providing one-on-one education. Then it solidifies what they learn by making them the teachers. In this way, students make massive gains in their educational successes.

Lessons in Free Market Principles

One last benefit of the Sabre program is worth mentioning. One of the best features of the Saber system is that it teaches students the basics of free markets.

In particular, the Sabre program teaches students the value of work, the value of consistent effort, and it gives them the understanding that they can better their situation through effort, education, and the workings of free market economics. Of course, they wouldn't articulate it this way, but that is exactly what they're learning.

THE CURITIBA EXPERIENCE

The city of Curitiba, Brazil had a problem[141]. It was growing rapidly. Most of the growth was in the form of shantytowns around the edges of the city with dwellings made of wood frames and cardboard. The lanes between the shanties were very narrow, so the garbage trucks couldn't get in. As a result, garbage began to pile up. As you would expect under such circumstances, disease broke out. There was no money in the city's budget for new streets, fresh water, and sewers.

---

[141] For further details, see Chapter 1 of New Money for a New World, Bernard Lietaer and Stephen Belgin, Qiterra Press, 2011.

Matching Supply to Demand

However, Curitiba did have assets. Curitiba was a lush area and so there was lots of cheap food available. Also, the city's bus system was underutilized because much of the population had no money for bus rides.

The city's mayor decided to use some very unconventional means to leverage the resources at hand to solve the problems that confronted the city. The city started giving out bus tokens to people in exchange for presorted, recyclable garbage. It also gave out plastic chits for paper and cartons. The plastic chits could be redeemed for inexpensive seasonal foods.

In essence, what the city government did was use bus tokens and plastic chits as complementary currencies to get the citizens of the shantytowns to clean the towns themselves. These currencies matched unused supply with unfilled demand in a way that solved massive social problems without taxes, transferring wealth, or going into debt. And it was all accomplished through custom-designed currencies.

The best part of all was that the people involved were performing needed and useful work to gain the rewards of the system.

Customized Money can be a Real Solution

The program was so successful that they decided to expand it. The schools had kids bring in cardboard and paper from the shantytowns and exchanged them for school supplies and school lunches. This was important because the lack of school supplies was what was keeping many kids out of school. Also, the lunches that kids got through the program were often the only real meal they had during the day. The kids responded by picking the neighborhoods clean. On the whole, the city's recycling efforts are estimated to have saved 1200 trees PER DAY.

People in the shantytowns used the bus tokens to ride the busses downtown to find and commute to jobs. Employment among the general population went up extraordinarily. Over the course of a single generation, more currency-based programs were used to finance the restoration of buildings, create parks and other green areas, and build affordable housing.

Removing the garbage from the shantytowns increased health, decreased disease, and improved the quality of life.

The increased economic activity that resulted from the rise of employment created new jobs, decreased poverty, enabled more people to move out of the shantytowns, and increased the tax base.

With food and school supplies available to even the poorest kids, the city improved educational levels, improved upward mobility, and decreased crime.

Both this and the previous examples demonstrate the power of social-purpose currencies. So now let's take a look at how we might be able to solve similar social problems with the blockchain.

THOUGHT EXPERIMENT: A PATH TO A BETTER HEALTH CARE SYSTEM?

Let's do a thought experiment that might just lead us to a better health care system than we have now. To start this experiment, let's first look at the problem that the health care system had before the Affordable Care Act, aka Obamacare.

THE ORIGINAL PROBLEM

What was the problem that Obamacare was designed to solve?

It was basically this: Not everyone could afford health care.

"So what?" you may ask. "It's always been that way."

True. But there were complications. By law, health care providers couldn't turn anyone away from their emergency rooms even if they couldn't pay. So people who needed health care went to the emergency rooms and got free care that way.

Government Intervention

Laws forbidding health care providers from turning people away were yet another government intervention in the free market that backfired. These laws made sure everyone had access to health care. But it came at a steep price. You and I had to fund the emergency room visits of those who couldn't pay.

You may say, "I don't mind. If people have emergencies and can't pay, I'll help out where I can."

That's very nice of you. But unfortunately, most of the emergency room visits by people who couldn't pay weren't emergencies. People who couldn't afford health care or didn't want to pay for it went to the emergency room for any health care need they had. So there were *lots* of extra emergency room visits that weren't emergencies and didn't need to happen.

This was driving up costs. But it wasn't the biggest factor in determining health care prices.

Another problem that was driving up the prices of health care insurance also stemmed from another government interference in the free market. For whatever reason, states routinely enacted protectionist legislation that prevented health insurance companies from selling policies to people in states where they didn't operate directly. In other words, if I lived in Maine, I couldn't get my health insurance from a company in Oregon even if the company in Oregon offered a better plan at a better price.

Protectionist legislation virtually always backfires, and so it was in the case of health insurance. The lack of competition within states was a huge reason for high prices and poor customer service. The government basically shot consumers in the foot and created an artificial cash flow for the insurance companies.

Of course, there were other things that drove up costs too. One of them was the numerous lawsuits that Americans hit their doctors and hospitals with. But because the lawsuits have their roots in legal rather than economic causes, I won't deal with them here.

The Real Problem

Although both of the issues presented so far were contributors to the original health care mess, they weren't the real problem. The real problems was that the cost of health care was masked from the consumer.

What does that mean?

People in the US mostly get their health insurance through their employer. That means that they never really see the health care bill. They don't have any idea what the hospital is charging them for their care. As a result, the hospital does its best to inflate the bill.

If the customers got the bill and had to pay directly, you'd better believe that they wouldn't put up with that. They'd go to another hospital and eventually the hospital that inflated its bills would be out of business.

In the free market, the price of something is an important piece of information. If customers don't have to pay directly for what they're consuming, they don't see the price. That leads to distortions in the market. And by that I mean that it drives the price up.

Health Care and Health Insurance are NOT the Same

To see things are they really are in the health care industry, we have to understand that health care does not equal health insurance. Just having health insurance does not mean you will get health care. And not having health insurance doesn't mean you won't get health care. You can absolutely get health care even if you don't have insurance. Health insurance and health care are just not the same thing.

Currently, we are used to paying for health care with health insurance. But that doesn't mean that that's the only way or even the best way to pay.

First, paying with insurance virtually guarantees price distortions. For example, suppose my wife and I both work for different companies. Now imagine that we both need the same procedure performed by the same doctor at the same hospital. The different insurance companies work out different deals with their provider networks. So she and I can both go to the same hospital and have the same procedure done by the same doctor and our insurance companies will be billed different prices.

"So what?" you may ask.

Well that means it's virtually impossible for my wife and I to shop around for the best deal. That's what we do when we buy everything else. If we're buying a new car, or a new couch, or almost anything else, we shop around for the best deal. That price information is tremendously important to us.

But in the health care system, using insurance to pay always distorts the price. We no longer care about getting the best deal for our money because we're not paying directly. So we just go where we like and let the insurance company worry about the bill. Because insurance disconnects us from the price information and distorts the price, it's virtually guaranteed that our bill will be higher than if we shopped for health care the way we shop for cars.

And it gets worse.

Because consumers don't shop around for health care the way they do for everything else, the insurance company must do it. They hire armies of people to go through past bills and reject billed items. It doesn't really matter whether the billed item was justified or not. If the insurance company thinks they can get away with not paying the hospital, they reject billed item.

The hospital reacts by employing armies of people that justify the rejected item or armies of lawyers that threaten to sue the insurance company. This back and forth between the insurance companies and the hospitals is a tremendous driver of increasing costs.

The System was an 85% Success

For all its faults, the American health care system worked and worked well. Fully 85% of Americans were covered. A portion of the remaining 15% just plain didn't want coverage because they were normally healthy. It was mostly young adults that fell into this category.

A system that works 85% of the time is a fairly good system in spite of its shortcomings. But that wasn't good enough for the politicians. Wanting to give handouts in exchange for votes, they decided to scrap the whole system and experiment with something totally untried.

Enter Obamacare.

OBAMACARE: WHY IT CAN'T WORK

As I mentioned previously, a lot of people think that having health insurance is the same thing as having health care. It fact, it's so ingrained in our thinking that the so-called Affordable Care Act is based on the idea that health care must be provided through a cumbersome insurance system plus a dizzying array of massive government bureaucracies. How can this possibly reduce the cost of health care and increase the efficiency of providing it? The whole effort was doomed from the start.

And that doesn't even take into account the fact that the primary web site for the Affordable Care Act, aka Obamacare, didn't work on the first day Obamacare was suppose to take effect. That web site, which was built by a company owned by a close friend of Michelle Obama, received more than $600 million for building a web site that didn't work at all[142].

Obamacare is like any government solution. It has completely backfired. Instead of being affordable, it has increased the cost of health care by an average of $3500 per family instead of reducing by a promised $2500 per family[143]. So the word "Affordable" in the law's title is a complete joke.

Obamacare is designed and built on the assumption that the only possible way to provide health care for people is if they buy insurance. It didn't occur to Mr. Obama or the Congresspeople who voted for Obamacare that there might be another way. And now there *can't* be another way. Obamacare *forces* everyone to buy insurance. This has resulted in windfall profits for insurance companies. But more importantly, it has completely locked out any innovations in the health care system that might make it more affordable.

> "President Obama has continually put off the deadline for implementation of Obamacare thanks to hang-ups in the system." - Ben Shapiro

Central Control Never Works

Being a centrally controlled system, Obamacare can't possibly work. The massive new horde of bureaucrats needed to run the new system is both centralizing government control over healthcare and driving up costs. There are a myriad of new boards, committees, regulators that have to be satisfied. Doctors, hospitals, and other health care providers are groaning under the burden of a mountain of new regulations.

All of this was completely predictable. And many did predict it. But they weren't listened to. People who voted for health care reform were stunned when they found that they would actually be paying for the very thing they voted for[144]. Apparently, they assumed "the rich" would pay, not realizing that the government defines "the rich" as practically everyone when it needs to.

---

[142] It was all wasted money. I would have gladly built the government a web site that didn't work for half of what they spent on the Obamacare site.

[143] Sally Pipes, "Obamacare Guarantees Higher Health Insurance Premiums -- $3,000+ Higher" Forbes, January 7, 2013, http://www.forbes.com/sites/sallypipes/2013/01/07/obamacare-guarantees-higher-health-insurance-premiums-3000-higher/ and David Hogberg, Ph.D, "Obamacare Exchanges: Less Choice, Higher Prices", National Center for Public Policy Research, March 2014, http://www.nationalcenter.org/NPA656.html.

[144] Jason Howerton, "Obamacare Supporter: 'Of Course I Want People to Have Health Care, I Just Didn't Realize I Would Be the One Who Was Going to Pay for It Personally'", The Blaze, October 7, 2013, http://www.theblaze.com/stories/2013/10/07/obamacare-supporter-of-course-i-want-people-to-have-health-care-i-just-didnt-realize-i-would-be-the-one-who-was-going-to-pay-for-it/.

At the time it passed, most people were against Obamacare and still are. The only way it passed was because the people behind it lied about it. In fact, one of its primary architects, John Gruber, admitted multiple times[145] on video that American voters would have killed Obamacare if they knew what it was really about. In his mind, it was ok to lie to Americans to pass the ACA because of the "stupidity of the American voter" in Mr. Gruber's own words. Mr. Gruber tells us that we just don't know what's good for us. We needed people like Mr. Gruber to tell us what is best and how we should live our lives, even if they have to lie to us to do it.

When Obamacare rolled out, the main web sites had such massive problems that they were completely non-functional[146]. Astoundingly, President Obama was surprised by this. Apparently no one actually tested the web site and he never asked about how it was going (or possibly his staff lied to him). Enrollment was much lower the levels that were originally targeted[147], and it has continued to be modest since then. People are not signing up for Obamacare in droves, especially among the young. Apparently, healthy young adults who don't make much money would rather pay the penalty than pay for Obamacare. It seems to be the cheaper option for them.

To date, the majority of the Obamacare exchanges–the web sites where people get their coverage–have already gone bankrupt[148]. No one in the Obama administration could see the very obvious fact that Obamacare was completely unsustainable from the start.

Because the entire design of the system was completely broken from the beginning, President Obama repeatedly issued illegal executive orders to delay its implementation[149], even though he had no authority to do so.

From the very beginning, Obamacare has been a case study in why central government control never works.

---

[145] Patrick Howley, "Obamacare Architect: Lack of Transparency Was Key Because 'Stupidity Of The American Voter' Would Have Killed Obamacare", The Daily Caller, November, 9, 2014, http://dailycaller.com/2014/11/09/obamacare-architect-lack-of-transparency-was-key-because-stupidity-of-the-american-voter-would-have-killed-obamacare/

[146] Jane, C. Timm and Traci G. Lee, "Millions log on for Obamacare, crashing sites", MSNBC, October 3, 2013, http://www.msnbc.com/msnbc/millions-log-obamacare-crashing-sites.

[147] Sam Baker," It's Official: Obamacare Enrollment Is Super Low", National Journal, November 13, 2013, http://www.nationaljournal.com/health-care/it-s-official-obamacare-enrollment-is-super-low-20131113.

[148] Tom Howell Jr., "More than half of Obamacare co-ops fail", The Washington Times, November 3, 2015, http://www.washingtontimes.com/news/2015/nov/3/more-half-obamacare-co-ops-go-belly/

[149] Alec Torres, "Read the Complete List of Obamacare Delays", National Review, The Corner, March 16, 20144, http://www.nationalreview.com/corner/374253/read-complete-list-obamacare-delays-alec-torres.

More Taxes!

Although most Americans aren't aware of it, Obamacare has increased nearly everyone's taxes. But the increases are cleverly hidden so most people won't notice. The new taxes are as follows[150].

- A 3.8% surtax on investment income when you make more than $200,000.

- A 0.9% surtax on Medicare taxes if your income is $200,000 or more.

- Your flexible spending account (FSA) is capped at $2,500.

- A higher deductible for itemized expenses (was $7,500, is now $10,000)

- The penalty on non-medical withdrawals doubled from 10% to 20%.

- A 40% tax on "Cadillac" health care plans.

- You can no longer buy over the counter drugs with your FSA.

- A penalty tax of 2.5% of your income if you don't buy insurance.

- A special tax on medical devices costing more than $100.

In what way does any of this make health care more affordable?

If You Like Your Plan, You CAN'T Keep It

To add insult to injury, the Affordable Care Act, which was supposed to get everyone in America covered by insurance, got literally millions of people kicked off the plans that they had and liked. And this was in spite of President Obama's now-infamous "If you like your plan you can keep it. Period." promise.

When people first started getting notices that their plans were canceled due to Obamacare, there were initial denials. But the cancellations continued until no one could pretend that the obvious was not happening. The number of cancellations exceeded the total populations of the states of California (38 million), Texas (25 million), and Florida (20 million) combined. Obama administration officials themselves later admitted that as many as 93 million

---

[150] Henry Blodget, "Here Are The New Taxes You're Going To Pay For Obamacare", Yahoo! Finance, Daily Ticker, July 2, 2012, http://finance.yahoo.com/blogs/daily-ticker/taxes-going-pay-pay-obamacare-145413745.html.

Americans would lose their health care plans[151]. After that first and very massive wave of losses, the Congressional Budget Office now says that Obamacare will cause *yet another* 10 million people to lose their plans. It just doesn't seem to end[152].

And because of centralized government control, the plans that Obamacare offers are fewer and less tailored to the needs of customers. This is why administration officials tried to force birth control on a group of nuns, The Little Sisters of the Poor. The nuns sued and the case went to the Supreme Court where the result was a resounding defeat of the Obama administration's legal position[153].

It's surprising that the Obama administration now has the audacity to claim that because of Obamacare, more people are covered than ever before. *Of course* more people are covered. It's *against the law* for them not to be. But their coverage is more expensive, less tailored to their needs, and comes at a cost of increased taxes. How does that make sense to President Obama and his political cohorts?

GOODBYE PRIVACY

In spite of HIPA privacy laws, Obamacare is a privacy nightmare. The government's health insurance web site actually has connections embedded inside it that enable third party tech firms to tell when you personally are on the site. Some can even pull out critical information like your age, income, ZIP code, whether or not you smoke, and whether you are pregnant[154].

---

[151] Avik Roy, "Obama Officials In 2010: 93 Million Americans Will Be Unable To Keep Their Health Plans Under Obamacare", Forbes, October 31, 2013, http://www.forbes.com/sites/theapothecary/2013/10/31/obama-officials-in-2010-93-million-americans-will-be-unable-to-keep-their-health-plans-under-obamacare/ and Robert F. Graboyes, "If You Like Your Plan, You Still Can't Keep It", US News and World Report, September 22, 2014, http://www.usnews.com/opinion/economic-intelligence/2014/09/22/under-obamacare-americans-will-continue-to-lose-coverage.
[152] "CBO Now Says 10 Mil Will Lose Employer Health Plans Under ObamaCare", Investor's Business, Daily, January 27, 2015, http://news.investors.com/ibd-editorials-obama-care/012715-736559-cbo-says-obamacare-will-push-10-million-of-employer-plans.htm.
[153] Robert Barnes, "Supreme Court says nuns are exempt for now from Obamacare contraceptives rule", The Washington Post, January 2, 2014, http://www.washingtonpost.com/politics/supreme-court-says-nuns-are-exempt-for-now-from-obamacare-contraceptives-rule/2014/01/24/12c7b9a0-853e-11e3-8099-9181471f7aaf_story.html.
[154] "Third-party connections prompt more privacy concerns about ObamaCare site", Fox News, January 20, 2015, http://www.foxnews.com/politics/2015/01/20/third-party-connections-prompt-more-privacy-concerns-about-obamacare-site/.

Your personal information also gets passed on to the IRS. Why is that a problem? This is the same government agency that admitted to targeting non-profit groups based on their religious and political affiliations[155]. When the whole illegal, immoral, and sordid affair quite rightly became a major scandal, the IRS promptly "lost" tens of thousands of the primary perpetrator's emails because the hard drive of the perpetrator, Lois Lerner, supposedly failed[156]. Apparently when you're in the government, subpoenas cause hard drive failures.

Are these the people we want to trust with our medical information? Are there ways in which the IRS, with its proven track record of going after groups the current political administration doesn't like, might use our health care information against us? I pity the person that finds out from firsthand experience.

HEALTHBUCKS

With the unending disaster that is Obamacare continually unfolding its nightmare of government central control before our eyes, we are certainly justified in asking if there is a better way to provide health care for Americans.

The answer is yes. There are many better ways of providing health care. But in order to enable the free market to innovate and solve this mess, we have to get the government out of the business of regulating the health care industry.

The next step is to devise a system that reconnects consumers to the prices of the health care they receive that is totally voluntary and applies free market principles to the problem.

Here's where we actually begin our thought experiment.

Let's suppose we have a blockchain-based currency that enables us to create a side chain. This side chain will issue a token that can only be used to pay for health care services. For lack of a better name, we'll call our token HealthBucks. But remember, this is not really money. The only people and companies that can accept HealthBucks are health care providers. And

---

[155] Patrick Howley, "BOMBSHELL REPORT: IRS Targeted 'Icky' Conservative Groups", The Daily Caller, December 22, 2014, http://dailycaller.com/2014/12/22/bombshell-report-irs-targeted-icky-conservative-groups/ and Stephen Dinan, "Judge orders IRS to release list of tea party groups targeted for scrutiny", The Washington Times, April 2, 2015, http://www.washingtontimes.com/news/2015/apr/2/irs-ordered-to-release-list-of-targeted-tea-party-/.
[156] If you really believe her hard drive failed, I've got a nice bridge in Brooklyn I'd like to sell you.

because the government loves to meddle in health care so much, we'll even say that it's against federal law to spend HealthBucks on anything other than health care.

For simplicity in regulating the supply of tokens, we'll say that HealthBucks is a mined "currency" that produces tokens at a steady rate according to a known algorithm. Miners mine the tokens and sell them on the main blockchain system's built-in currency exchange.

And because we're involving the government, we'll even say that customers can purchase HealthBucks with pre-tax money. That is, the money they spend on HealthBucks is not taxable income. So it saves them money.

Don't these concessions to government meddling mean that the government is picking winners and losers again?

Possibly, but not if they make the same concession to other approaches to providing health care.

Remember that I'm not saying that HealthBucks should be the one and only way to get health care. I'm saying it could be one of many possible solutions that consumers can choose from. So as long as the government is willing to give the same considerations to HealthBucks that it gives to the insurance industry–or anyone else–it shouldn't be too much of a problem. That being said, if you can come up with a better implementation, please create it and I'll use it.

How Would this Be Implemented?

So far, all we've got is a "currency" that's not really money. How does this solve our problem?

Plans for Everyone

What if you had a system that you could pay into on a monthly basis and receive a set amount of health care for that payment? In other words, for $500 per month, you get Plan A, which gives you X number of checkups, visits, emergency care, and so forth. For $1000 per month, you get Plan B, which gives you Y number of checkups, visits, and so forth. There can be as many plans as the health care provider wants to offer.

Why would this be better? Because it's the health care providers themselves that offer the plans, not the insurance companies. You're paying your doctor or hospital directly rather than involving the insurance companies. The insurance companies are totally out of the picture. This makes providing you with health care *much* cheaper for doctors and hospitals. So they can pass that savings on to you in the form of much cheaper health care.

Plans as Smart Contracts

The blockchain actually provides multiple possible approaches to implementing the HealthBucks system. You can take many approaches once you have the HealthBucks side chain built. Probably the easiest way is with smart contracts.

Each doctor, hospital, dentist, and other health care provider can create their own smart contracts. These can be standardized contracts or it can be left up to each health care provider to customize their plans the way they want to. Most likely an industry would spring up that provided customizable contracts cheaply.

To sign up for health care, customers get on the provider's web site and use it to create the contract. Once the web site created the contract, it would sign the contract on behalf of the health care provider. The customer must sign it too. Since both parties have signed the contract digitally, that means there's a legally-enforceable agreement between them. The doctor, hospital, dentist, or other health care provider then gives the health care that the patient requests and the patient pays using HealthBucks.

Using HealthBucks

Each month, the customer pays money into the contract. The contract automatically deposits the money in an account owned by the health care provider. It then gives the customer HealthBucks. The HealthBucks just appear in the customer's digital wallet along with their other digital currencies.

When the patient arrives at the provider's office, they pay for any services that their plan allows using the HealthBucks. Patients can spend their HealthBucks in any way they want. If a patient wants to spend the majority of their HealthBucks that month on dental work, they can. If they need foot surgery the next month, their HealthBucks can be spent there.

When the HealthBucks are gone, the patient has used up the health care that they've purchased. This eliminates the need for health care providers to have any sort of accounting system at all. They get it for free through the blockchain. As a result, their costs go down and they can pass that savings along to you.

Emergencies

For extreme emergencies, patients who do not have sufficient HealthBucks could purchase catastrophic health care insurance or self-insure. But if the system lets them save up their HealthBucks from month to month, then patients accumulate their HealthBucks when they are healthy. This enables them to quickly get to the point where they can pay for everything but the most dire emergencies on their own.

Recall that you can buy HealthBucks and you can spend HealthBucks. What the system notably lacks at this point is a way to turn HealthBucks into a currency like dollars. No one can convert HealthBucks to money except miners and health care providers. That is the way the system is designed. It keeps people from using their pre-tax money for anything other than health care.

We can make the enforcement of the limitation on the redemption of HealthBucks for money easier by saying that only health care providers can be miners. Again, we're bowing to those who love government control. But this isn't a serious limitation in terms of free market principles because tech companies can provide mining services to health care providers. That's where the free market competition comes into play. Anyone can create a company that builds mining computers (they're different that regular computers) and performs mining services for doctors and hospitals. The health care providers pay for the mining services with digital cash or dollars. The companies providing the mining services never see any HealthBucks. The HealthBucks go directly into wallets owned by the health care providers. And only health care providers can redeem the HealthBucks for money.

Changing HealthBucks to Money

How do the doctors and other health care providers get money in exchange for the HealthBucks that patients pay them? After all, it isn't HealthBucks they want, it's money like digital currencies or US dollars.

Health care providers can use the blockchain system's built-in, distributed, decentralized currency exchange to sell their HealthBucks. The HealthBucks are purchased by the smart contracts that drive the system.

Remember the smart contracts? They automatically get onto the blockchain's currency exchange and use the money that customers have paid into the contract to buy HealthBucks. That's how they get HealthBucks to give to customers. And when the smart contracts use money, usually digital cash, to buy HealthBucks, the health care providers get their money.

Accumulating HealthBucks

Taking this idea further, shouldn't we reward families with healthy lifestyles? If we could encourage people to exercise, eat in a healthy way, quit smoking, lose weight, and so forth, it would significantly reduce the costs of our health care system. So how could the subscription system provide these incentives?

One way to do that is to let patients accumulate HealthBucks from month to month and year to year. That is, if my family and I live a healthy lifestyle, we should be able to keep any HealthBucks that we don't spend each year. In fact, we should be able to pass that down tax-free to our dependents. This would encourage families to raise children with a healthy lifestyle.

Parents can save HealthBucks for their declining years. Or they can accumulate enough of them so that they don't need catastrophic insurance any more. They are self-insured. They can also spend HealthBucks on their children or grandchildren no matter what age they are, so families can share resources and cushion themselves from catastrophic illnesses or accidents.

What's the Advantage?

There are distinct advantages of the HealthBucks approach over what we are doing now.

Savings for Providers

As I've already stated, this approach is much less expensive for health care providers because they don't need to deal with insurance companies. Billing insurance companies and fighting with them over the items billed costs health care providers massive amounts of money each year. Remember that this is one of the big reasons that health care costs were getting out of control in the first place.

Providers can encourage customers to use this subscription system by passing their savings along to their customers. In other words, providers charge patients in the subscription system far less than patients paying with insurance because it's cheaper for doctors to use the subscription system.

Also, we saw already that the providers don't need to create a special accounting system for subscription patients. It's built into the blockchain system. Patients' spending is controlled by the amount of HealthBucks they have. So that's what they can afford. If they have the HealthBucks and the expense is allowed by their plan, then they simply spend their HealthBucks in the way they want and the doctors don't have to track it. The patient's wallet keeps track for them.

Savings for Patients

For the average consumer, the subscription approach is cheaper and easier. They know exactly what they can get from their doctor. And even with the cost of the subscription system plus catastrophic health insurance, the average consumer would pay less per month than they are paying now.

Eventually, most patients would be self-insured and the overall costs of the entire system would be reduced. In addition, the ability to pass your accumulated HealthBucks down to your descendants tax-free would reward thrifty families that make the effort to save for a rainy day. So for example, if one of your grandchildren was born with a birth defect that required expensive health care for long periods of time, you could help insulate that child's parents from large health care bills by sharing some of your accumulated HealthBucks.

And again, there's always catastrophic health insurance for extreme emergencies. Most health care providers would try to make their plans more attractive by simply making bulk buying deals with the insurance companies. That way the providers could offer the their plans plus catastrophic health insurance at a reduced price. The health care providers could use their large customer base to get a discount on insurance for their patients.

The health insurance companies would make money because they would have all of the patients of particular doctors buying catastrophic health insurance from them. That's a good revenue flow. And given that most doctors tend to form themselves into large health care provider networks, the potential business in catastrophic coverage could become very lucrative without becoming a burden to patients.

The patients save money because they're getting complete coverage at a discount. And it's coverage that they choose, not coverage that the government chooses for them.

Market forces would also help keep the costs of health care down. The providers would want to offer good plans at an attractive price. Because all of the providers would be competing for HealthBucks, they would necessarily need to offer the best and most cost-effective health care possible. The result would be a huge advantage to the average customer.

One Size Fits All?

Will this subscription-based system work? We think it will certainly work better than the current system, which is collapsing around us. It will certainly work for the vast majority of Americans.

Will this subscription-based system work for *absolutely everyone*? Probably not. For people with chronic illnesses or families with heavy health care needs, this system may simply not work. But that's the nature of freedom, free choice, and free markets. There is no one-size-fits-all solution because one-size-fits-all in reality means one-size-fits-none. Any attempt to create a single, universal solution for absolutely all health care needs is doomed to failure. There is no single, comprehensive system that can be thought up by any single person or central government planning committee that will work for absolutely everyone. It's just not possible.

Instead, we can use the blockchain to create an infrastructure that enables anyone to innovate on health care systems. This lets people have the freedom to create what they think is best and choose what they think is best. Freedom solves problems that governments can't. In this way, new health care systems will emerge that meet the needs of all customers. We know this because this is what happens in all other markets that the government doesn't heavily regulate. The more government regulates health care, the poorer health care system we'll have.

The Last Resort

For those that absolutely can't afford any health care at all, we can implement a need-based system that gives them HealthBucks. I personally don't like this solution because it involves the government again. But it is something that I think will please the statist, big government worshippers without compromising the free market too badly.

Basically, there would be a fund that anyone can donate to in exchange for a tax deduction. This would make it non-compulsory, so no one is forced to donate. Just as we send out kids trick-or-treating for UNICEF, people can freely solicit contributions to this fund. But no one would *have* to pay.

People who need health care but are unable to pay for it can submit their tax returns to the IRS for an audit. Yes, I know, we all hate the IRS. But in this case, we're turning them into a money-giving organization rather than a money-taking organization.

So people apply to the charitable HealthBucks fund and submit their tax returns to the IRS. The IRS validates that they're qualified to receive funding from the charity.

Now this has the potential to really drive up the costs of the system over what I've outlined so far. And it's one big reason why I don't like the solution. Still, it's better than what we have.

Once the IRS declares people qualified, they'll automatically receive a set amount of HealthBucks in their digital wallets that they can save for when they need health care.

As I stated, I really don't care for this solution of last resort. I'm sure that someone out there can think of something that's better, simpler, and cheaper. That's the nature of the free market. It applies more brainpower to a problem than any individual or any government agency can supply.

ANYONE CAN CREATE A SOCIAL-PURPOSE CURRENCY

So with these examples of how social-purpose digital currencies can solve massive problems, it's easy to see why building a blockchain-based toolset is a highly desirable approach. It enables any individual, company, nonprofit organization, or governmental body to create a social-purpose currency to solve otherwise intractable problems.

Money is not value neutral. People spend it differently depending on its features. Applying free market principles to social problems is a matter of matching a supply of unused resources to unmet demand. Currencies whose features connect supply to demand in ways that are not done now will be able to impact the behavior of anyone using that currency. If the currency is properly designed, then the impact can be very positive indeed.

By applying free market principles to money, we create incentives but use no force. In the HealthBucks example, there was no element of force or compulsion in any part of the system. Everyone had financial rewards for participating. They were all free to choose to use HealthBucks or not as they saw fit. And that freedom to choose is the essence of the free market.

13 THE BLOCKCHAIN AND THE BUSINESS REVOLUTION

As you might expect, the blockchain has profound implications for business. It will fundamentally change many aspects of how our economy works. In this chapter, I'll present just a few possibilities that the blockchain offers for business applications.

COMMERCIAL CREDIT CIRCUITS

In the 2008-2009 financial meltdown, we witnessed the brittleness of our economy. One reason that our economy is so brittle is that there are too few ways for small to medium sized business (SMBs) to obtain short-term credit when they need it to do business.

SHORT-TERM LOANS DRIVE MANY PARTS OF OUR ECONOMY

Suppose you run a company that manufactures machines for factories. Imagine that you get a large order to design, build, and install specialized machines for a factory. So you have a paying customer but you will not receive any money until you finish the job.

To complete the job, you need to get a bridge loan from the bank. A bridge loan is a short-term loan that you get from a bank to finance the project you're working on until the customer pays you. Bridge loans are common in a lot of manufacturing industries and in the construction industry.

Well getting a bridge lone is fine unless the banks are not making loans due to a difficult economy. In which case, bridge loans are not available so you essentially cannot do business.

When situations like this occur, the transaction cannot be completed because there are too few sources of liquidity for SMBs in our economy. So what do you do when you can't get needed short-term loans from banks?

The answer might be to use a commercial credit circuit (C3). Here's how it works.

Using Commercial Credit Circuits

To complete your order when the banks won't give you financing, you join a C3. Anyone with the right business know-how can start a C3, including banks. C3s are not hard to implement with the blockchain's smart contracts.

Joining a C3

When you join a C3, you take out insurance on your order. The insurance only pays off if the client that placed the order fails to pay upon delivery. Depending on the rules of your C3, you might buy the insurance from the C3 or obtain it at a local bank or insurance company. Banks are more likely to sell your company insurance on an order during hard times than to extend a loan because there is less risk for them. They are not putting out any money up front.

Your next step is to use the C3's server to issue a currency that is backed by your order and insured in case the client defaults. Smart contracts can create this currency out of thin air.

Your currency really is a set of cryptographically signed IOUs that have the same properties as a digital currency. So you can use the IOUs just like money.

Using Your C3 Currency

Next, you use the C3 currency to pay your suppliers and possibly your employees. The C3 currency that you are using is essentially an insured invoice and you use it to pay anyone who will accept it.

Anyone who holds the currency can cash it in for digital cash or dollars (or other national currency) at any time. However, they must pay an interest fee that is prorated to the date that the client is expected to provide payment. The earlier the currency is cashed in, the less it is worth. If currency holders wait until the client has paid, then they get the face value of the currency. Even if the client doesn't pay, they still get their value because of the insurance. Either way, they can redeem the C3 currency for cash because the C3 currency is really just a set of digital IOUs.

> "The blockchain's promise extends well beyond financial services; it extends into adjacent verticals such as healthcare and land rights. The common denominator that will link all digitally enabled services could very well be blockchain-enabled digital identity systems, securely stored and managed in a distributed ledger." — Menekse Gencer

What can suppliers and employees do with the currency you issue? If they also join the C3, they can use it to pay anyone else. Whoever they pay has the same choice; cash in or join the C3 and continue to use the currency.

Your supply chain is more likely to participate in the C3 than it is to participate in other forms of alternate currencies because of the insurance that you buy up front. They know for sure that one way or another they will get their money in the end.

COMMERCIAL CREDIT CIRCUITS IN THE ECONOMY

Because you do not need to go to a bank for your short-term loans, you can keep doing business using C3s even in difficult times.

Will C3s work?

Absolutely. They have already seen some use in particular industries and have developed a proven track record. The biggest hurdle is just convincing your suppliers to accept the C3 currency. But once that hurdle is cleared and businesses learn that they are guaranteed to get their money if they accept the C3 currency, you can use C3s repeatedly to do business on an ongoing basis.

> "So [the blockchain] is an extraordinary thing. An immutable, unhackable distributed database of digital assets. This is a platform for truth and it's a platform for trust. The implications are staggering, not just for the financial-services industry but also right across virtually every aspect of society." - Don Tapscott

Many industries can use C3s. Having C3s would have been particularly helpful during the start of the Great Recession. They would have softened the blow on the housing industry by providing a replacement for the construction loans that are commonly used to build houses.

The nice thing about C3 currencies is that you use them just like you use digital cash. You can store them in your wallet right along with your digital cash. The experience of paying with your C3 currency is just like paying with digital cash.

BRANDED CURRENCIES

Recall from previous examples that blockchain systems can be used to create branded currencies. These include customer loyalty programs, discount coupons, and so forth.

For instance, I previously presented an example in which a company called East End Publishers created a customer loyalty currency to encourage customers to continue to buy books from them. Blockchains, together with smart contracts, can be an easy and inexpensive way to set up customer loyalty programs, cashback programs, discount coupons, "Green Stamp"-type programs, and many more rewards systems that incentivize customer purchases.

The first company that builds a blockchain system that enables anyone, no matter how small their business is, to create branded currencies is the one that will score big in the market.

DIVIDEND-BASED CURRENCIES

The blockchain enables anyone to issue a coin that's used as a token. The issuer of the token can distribute it to whomever they want. Issuers can then use that token to distribute dividends to holders of their tokens.

For example, a musician can issue a branded coin for an album that he wants to finish and release. When fans buy the branded coins in exchange for digital cash, the musician uses the digital cash to pay for finishing the album. After the album is out and profits come in, the musician can pay dividends to everyone who holds his branded coins. In this way, fans can support independent musicians and receive financial rewards for doing so.

Dividend-based currencies can also be used to fund startup companies. A startup company that sells a branded, dividend-based currency can use the funds to get going and then distribute profits to supporters as dividends.

Using dividend-based currencies is, in many ways, like selling stock in a company or other endeavor. However, the currency usually does not grant ownership. Therefore, it is fundamentally different than a stock. Nevertheless, the ability of businesses to use the blockchain to create dividend-based currencies enables individuals and organizations to share profits with supporters in ways that are beneficial to both.

POWER TO THE PEOPLE: BUILD YOUR OWN MICROECONOMIES

One way out of the dilemmas created by the kind of money we use and our dysfunctional banking system is to transcend money entirely. In the process, we can democratize the economy by decentralizing control over it.

Transcending money is not as hard or as esoteric as it sounds. We simply use a process called *credit clearing*, which has been employed by banks for literally centuries.

WHAT IS CREDIT CLEARING?

The idea behind credit clearing is actually very simple. There's an example to illustrate it.

Suppose that you have a friend named Al that owns the corner diner. Al lets you put your meals on a tab and pay for them at the end of the month. So you sign an IOU for Al every time you eat.

Imagine that you keep several chickens on your property. You sell the eggs to another friend of yours named Bob. You let Bob put the eggs on a tab and pay for them at the end of the month. Bob signs an IOU every time he buys eggs.

Now let's say that the IOU Bob gave you for this month happens to be the same amount as the IOU that you gave Al for this month (you've been eating at Al's a lot). You, Al, and Bob are all high school buddies. Al knows Bob will pay his bills. So you just hand Bob's IOU to Al and Al cancels your debt to him.

In this transaction, no money has changed hands. You were able to both buy from Al and sell to Bob without any money at all.

This is the essence of credit clearing. All exchanges are done with IOUs.

If you scale this scenario up to hundreds or even thousands of people, you have a credit clearing cooperative.

Note

A credit clearing system IS NOT bartering. When you scale this up to the size of a cooperative, you use your IOUs like cash so you do not have to find someone whose needs align perfectly with what you produce, as was done in this limited example.

LOCAL FREE MARKET ECONOMIES

Credit clearing cooperatives enable people to band together with others that they trust, form their own free market economy, and innovate new economic instruments in any way they want. They can then join their cooperatives into large networks that provide all of the goods and services they need. And they can do all of this without any currency at all.

Credit clearing cooperatives encourage locality. In other words, it's unlikely that someone from Hong Kong will want to be a member of a co-op in Declo, Idaho. And it's unlikely that the folks in Declo will want their members scattered that far away. So the people in a given area will tend to form their co-op and run it themselves. This encourages people to buy locally and provides a reason for local businesses to join co-ops.

CO-OPS ARE BUSINESSES

Co-ops are businesses and they exist for the purpose of gaining profit for their members. When you join a co-op, you become one of the owners. This gives you an equal vote with everyone else in the co-op. It's another way to democratize the economy.

Co-ops should not be run on volunteer labor. Credit clearing co-ops run by volunteers have been tried in the past. Experience shows that after an initial burst of enthusiasm, they die out or decline rapidly. Therefore, the co-op should be run as a business (or nonprofit) with a paid staff.

Co-op businesses are not hard to implement using smart contracts. And the credits they produce can be made to automatically integrate with one or more blockchains. So if you have a stable blockchain currency that can hold its value no matter what's happening in the economy, you can use smart contracts to implement your co-op system on top of that blockchain. That way, your credits can be used in the blockchain's distributed currency exchange. They can also be incorporated into financial instruments and smart contracts based on that blockchain.

JOINING A CO-OP

In general, co-ops should have at least 200 members but not more than 5,000. Fewer members than 200 results in the co-op having an economy that is not diverse enough. Also, the co-op must make an income and having too few members usually means it can't make enough money.

Having more than 5,000 members in a co-op dilutes the vote of individual members and makes them feel disconnected from their economy. They will not want to participate much.

Anyone who joins a co-op does so by buying stock. This makes the member an owner. As an owner, every member has a vote in the co-op's policies. This provides members with greater control over their own microeconomy. It also decentralizes economic control away from national governments. In a true democracy (or democratic republic like the US[157] and many other countries), everyone should have a say in how their economy works. It should not be centrally planned or controlled at all.

Also, everyone should have a chance to compete in an open, fair marketplace. The current system not only favors Big Money and Big Government, it actively promotes their interests above the interests of the rest of us. In the co-op system, everyone competes according to the principles of the free market. And if you don't like how your co-op is operating, you can go join another one or start one of your own and compete just like everyone else.

To join a co-op, you must first find a co-op with people that you trust and rules that you like. If you can't find such a co-op, you can start one yourself.

When you join, you pay an entry fee. This fee buys you stock in the co-op, giving you the privileges of ownership.

You also undergo a background and credit check. Co-ops must ensure that their members are reliable. Of course, it's possible for normally reliable people to have personal problems and go bankrupt. In that case, the co-op can provide itself with various types of insurance to handle the situation. It

---

[157] We are a republic because we're governed by representatives. We're a democratic republic because the representatives are elected by a democratic vote rather than appointed by a king or some other entity.

can also provide remediation plans for members who are recovering from life's difficulties. But everyone in the co-op system must be reliable under normal circumstances. A few bad members can ruin the reputation of the whole co-op, so you need to screen your members carefully. In fact, you need to screen them as well as banks and credit unions do when they give out loans.

CO-OP CURRENCY THAT'S NOT MONEY

Instead of currency, all participants in a credit clearing co-op use certified IOUs, called *credits*, that they issue themselves using a blockchain-based system. Their "money" comes into being when they get ready to spend it. The system cancels their credits when they provide goods and services to the co-op system and accept credits as payment.

Legally, your IOUs are personal checks. But instead of redeeming your checks with US dollars or digital cash or any other type of money, you redeem them with goods that you create or services that you provide. So a credit clearing cooperative is a way of doing business with checks rather than money. But the checks are all digital and have the same characteristics as digital cash. So as you might expect, you can keep them in your digital wallet and use the blochain infrastructure to buy, sell, and do business.

Issuing Credits

A participant in the credit clearing system issues the credits themselves. It's just like writing your own personal check. Credits can only be spent between members of your particular co-op. They have no value outside the system so they are not considered a currency, they're just IOUs.

When you issue credits, you'll notice that they are in specific denominations just as if they were money. Credits are designed to simulate the experience of using money. But even though you see "bills" in your digital wallet of 5, 10, or 20 credits, no money exists. These are just IOUs made out in specific amounts. After you have issued yourself some credits you can buy goods and services from other members of the co-op just as if you were using money.

You issue credits against your credit limit, which the system keeps track of. As you buy and sell in the co-op system, you can use credits that you accumulate to increase your credit limit. Basically, you pay the co-op some of your credits and they increase your credit limit proportionately.

When you hit your credit limit, you have to stop spending. You can pay down your outstanding balance by selling goods and services to other co-op members and receiving their credits. You use their credits to pay down your balance. As you pay down your balance, you become able to issue new credits.

When you sell goods or services and accept credits from others, you can accumulate credits beyond what you need to pay off your balance. You can then you can save, invest, and spend the additional credits.

Credits are Private

Credits have all of the properties of digital cash. So like digital cash, credits are completely private. They do not identify the person who spends them. Even though you are spending IOUs, the IOUs are certified by your co-op. So your identity is not needed or tracked by the system when you make a purchase. All the system knows is how much you spend. It must know this so it can cancel your credits when you redeem them.

Also, the system doesn't keep track of your individual purchases. It only knows the total balance of your unredeemed credits, which is your unpaid balance. So it can't be used as a source of information about purchases you've made.

The only other piece of information the system tracks is your credit limit. It tracks this so that it knows when you can't issue any more credits. You have to redeem some credits before you can start spending again.

Reputation Score

The credits a buyer issues contain the buyer's reputation score (but not identity), which is assigned by the buyer's co-op. So when the buyer attempts to pay, the seller can decide whether or not to complete the transaction.

In addition, the credits contain the reputation score of your co-op. So although the buyer can remain anonymous, the seller always knows the reputation of the person he or she is dealing with as well as the reputation of the buyer's co-op.

The experience of using credits is specifically designed to seem like you're using a currency. Because it seems like everyone is issuing currency in the system, it's actually possible that a co-op or its network can name their credits as if it were a currency. That is, they can call their credits OurBucks (or whatever) and actually trademark the name. However, this does not imply that anyone in the system is issuing a currency. They are not.

CREDIT CLEARING CO-OP NETWORKS

No one co-op can provide all of the goods and services its members need. But that's ok. Co-ops can use smart contracts to band together and create a network of affiliated co-ops. This expands the amount of goods and services available to everyone.

If you buy something from a seller in an affiliated co-op, the seller can accept your credits just as if you were both in the same co-op. The co-op system automatically takes care of everything.

It's even possible for your entire network to join another network. The co-op system can be as large as people want it to be. But you still have local control because you are an owner in your co-op.

Co-op networks impose few rules on their affiliated co-ops. Those rules are mostly just the normal rules of business. So you don't really sacrifice any autonomy by joining a co-op network.

Affiliation networks audit their member co-ops on a regular basis. If their business practices follow the rules of the network, the network assigns the co-op a high reputation score. If not, the network may lower the co-op's reputation score.

Because networks are self-policing, no government intervention is needed anywhere in the system. If your co-op finds that the network they've joined is sloppy in enforcing good business practices among its members, then your entire co-op can go join another network. You have the freedom to leave an economy that you don't think is fair.

Co-op networks also provide investment services and banking services that your local co-op can't. For instance, your network may make a deal with a nationwide bank to offer mixed credit and dollar services.

As an example, suppose you go into a store owned by a member of your co-op's network. The store owner may or may not be in your particular co-op. But it doesn't matter. If the owner is in a co-op in your network, you can use your credits to pay for items in the store.

But the store owner must pay taxes on purchases even if they're in credits. So the store owner probably wants you to pay partially with credits and partially with dollars (or other national currency). Your network can cut a deal with a bank to enable that to happen. Your bank can provide you with a credit or debit card (or both) that lets you spend both dollars and credits with the same card.

So if the store owner wants you to pay with 60% credits and 40% US dollars, it won't matter. You swipe your card to pay just like you do now and the system will pay with 60% credits and 40% US dollars. For you, it's as easy as what you do right now. But this is only possible on the larger scale offered by a network. Individual co-ops usually aren't big enough to get this kind of a service from a bank.

So in short, your co-op network offers you access to a much larger economy without sacrificing local control. It gives you investment opportunities that a single co-op can't offer. It can also integrate the credit economy with the dollar economy by giving you integrated banking services.

But what if neither your local co-op nor the co-op network produces a good or service that you need? The answer is, of course, that you can use blockchain-based digital currencies or even your national currency to complete the transaction. Being in a co-op doesn't limit you to spending only credits. The availability of blockchain-based digital currencies and national currencies means that you always have a means of paying for what you need. The fact that your co-op is simply a business built with smart contracts that are based on a stable blockchain currency means that you have easy access to the wider economy. It also makes it very easy to network your co-op with other co-ops that are based on the same blockchain. You can even write smart contracts that handle the exchange rate between your credits and the credits issued by any other co-op.

WHY WOULD I JOIN A CREDIT CLEARING CO-OP?

At this point you're probably wondering why you would bother joining a credit clearing co-op when you can just use US dollars. Here are a few reasons why.

Decentralized Economies

The both US government and our centralized banking system are the sources of most problems that occur in our economy. They cause inflation, push people toward poverty no matter how well they manage their money, create inequity, and centralize control over all aspects of human life. Credit clearing co-ops and networks of affiliated co-ops enable you to create entire economies that are not under the control of central bankers and the government. They enable you to participate in a rebirth of the free market. They democratize the economy by spreading control out to all who participate in it.

Freedom from Inflation

Credits come into existence when a member issues them. They are redeemed when the member sells goods or services to other co-op members. As a result, there is never any inflation or deflation. The number of credits in use can never be larger or smaller than the total value of the goods and services in the system.

The fact that credits in circulation exactly matches the value of the goods and services available means that credits are a more robust form of exchange than any type of money currently in use. If you denominate your co-op credits in terms of a stable blockchain currency that never inflates or deflates, both your credits and your money will have a stability that national, government fiat, debt-based currencies can't possibly provide.

New Investment Opportunities

Your credit clearing co-op system can define financial instruments that are not based on the collection of interest. Being able to invent your own financial instruments to replace such common things as mortgages and other financial vehicles enables you to create new investment opportunities that you do not have now.

Because your co-op is a group of people that you know and trust, it can invest in ways that individuals can't. Essentially, you bring the power of the corporation down to the individual level. As a group, you and your fellow co-op members can decide what kinds of investments you want to create and then invest together. You can build co-op owned businesses, make loans to members, and so forth. Or you can create a co-op owned C3, start a lending club, offer unmortgages, and anything else you think will make you money.

Having a good portfolio of investments can enable a co-op to do much more than just handle financial transactions and investments. For example, a co-op or a group of co-ops could decide to start a private school. Of course, the tuition for nonmembers would be higher than the tuition for members. Members may find that the co-op system makes it easy to send their children to private school. Because members are the owners of the school, they have a more direct say in what is taught than is available through public schools. And if the co-op's school proves popular, it can provide a good return on the investment.

Co-ops can take this same approach with other services. They might want to use investment groups to start a health care clinic or hospital. This enables the co-op system to provide its members with affordable health care that is directly responsible to the patients that it serves.

In addition, co-ops can use investment groups for long-term retirement planning. Smart investments that produce steady revenue streams can eliminate the need to depend on meager government-funded retirement plans.

You Own Your Economy, So You Decide the Rules

As a member, you are an owner of your co-op. As such, you have a vote in all of its policies. Your co-op network provides financial services, offers investment opportunities, and audits your co-op to ensure that your co-op is using fair business practices. Other than that, it does not impose any rules to speak of upon your co-op.

So you get to decide what happens to the stock members own when they divorce or die. Most co-ops will probably follow existing procedures that are normally used for stocks and other assets. But if your co-op wants to add special provisions, it can.

You get to decide whether or not your members must be local or if they can keep their membership when they move away. Some co-ops may ask the members to sell back their stock when the members move out of the local area. Others won't care. Some co-ops may change a member's status to a non-voting membership if they live outside the local area. Whatever seems right to you, you can do. That's what freedom of choice is all about.

You decide whether or not you'll allow children to inherit their parents' memberships and credits. And if you don't like the rules that a particular co-op has on this subject, you can just join a co-op whose policies you like. Or you can start your own.

As previously stated, co-op members can pool their savings and invest together. What investments should your co-op offer its members? You decide. Your vote counts as much as anybody's.

Which affiliation network should your co-op join? Here again, your vote counts.

This system is about freedom, and freedom means that you get to decide how your local microeconomy works.

CO-OPS AND BANKS

Because co-ops engage in functions that are bank-like in many respects, it would be natural for co-ops or co-op networks to want to start their own banks. A simple way to do this is for them to create a credit union. If they did, they could freely mix their financial activities between credits and US dollars (or other national currency).

Existing banks can see this as a threat or a business opportunity. A forward-thinking bank could see this as a chance to get into new markets. For example, imagine that there is a nationwide bank called the First National Bank of Elberta. First National could provide all of the banking services that a co-op or co-op network needs if they would just accommodate credits as well as dollars.

To move into this new market, First National can offer co-op networks services such as transaction processing, savings and checking accounts in credits, and investment group management.

In addition, First National could provide interface services between the credit and the dollar economies as we saw in previous examples. They could enable people to pay in a mixture of dollars and credits. If they were smart, they'd offer debit cards in credits.

In such an arrangement, First National would be able to offer its services both online and at branches nationwide. So members could withdraw credits at any branch nationwide. Co-op members might also be able to obtain credits via ATMs that you store on cards with magnetic strips.

Also, many banks sell business-related insurance policies. Every co-op would need some form of insurance to protect itself against members who default on their debts. This is a chance for banks to get into a market that they do not currently serve.

As you can see, there are many opportunities here for banks to make money. If they simply add their services to the credit economy, they can assure themselves of new sources of revenue.

BUSINESS, BUSINESS EVERYWHERE

The blockchain can streamline business processes in ways that are not possible now. Smart contracts and side chains are the building blocks of new economic infrastructures. These new innovations can help us build a more robust economy and make money doing it.

Commercial credit circuits enable businesses to get the short-term liquidity they need to keep operating through good times and bad. These are easy to implement through smart contracts and side chains.

Likewise branded currencies provide an inexpensive way of creating customer loyalty programs and other special-purpose negotiable financial instruments. A simple side chain is all it takes.

Dividend-based currencies enable new investment models that can provide startups with needed funds for building their businesses. They can be an easy way to share profits with investors.

Finally, you can build your own microeconomy that helps you boost local businesses, invest in your community, and free yourself from government-run economies. Smart contracts and side chains make this relatively straightforward.

14 EPILOG: THE FREE MARKET IS AWESOME

You've made it to the last chapter. Congratulations. Along the way, we've talked about the serious and daunting problems our civilization faces. The fact that these problems could only be touched on in limited ways in this book should not minimize the significance of them. It's not an exaggeration to say that we're standing on the edge of a precipice. And a wrong move or two really could destroy our civilization.

We've seen that the real root causes of most of our world's problems come from the kind of money we use and the banking system that we depend on. Both our money and our banking system are detriments to the sustainability of our world.

Fortunately, freedom gives us choice. The free market–people choosing to use their resources in the way they think is best–provides us with ways to innovate our way out of many of our difficult situations.

We can provide ourselves with currencies. And we can do it better than the government. The free market lets us build new economies that are fairer, more efficient in many ways, and more robust than what we have now. By applying free market principles to money itself, we can construct fairer, more humane economic infrastructures than have ever existed in human history.

Let me reiterate that the free market cannot solve absolutely all of our problems. But it sure can make most of them better.

There are certain basic human rights. One is the right to life. Close to that is the right to reproduce. And after that is the right to be free. Implicit in that is the right to be left alone.

The free market is based on and reinforces basic freedoms and rights, especially those delineated in the US Constitution. Without the free market, power concentrates from the many to the few–because control over economic activity is control over almost all aspects of human life.

The free market in action is the greatest engine of prosperity that has ever been invented. And we can use blockchains to create the purest implementation of the free market that ever existed. We can build economies that are not centrally controlled or subject to political machinations. We can decentralize power, provide a fair market, create more humane economies, and offer people real pathways out of poverty.

The most important legacies that we can leave our children are things like stable money to base their economy on, a fair and humane chance for everyone who will work to get ahead and make a better life for themselves, and the freedom to make the future what they want it to be.

Those legacies are precisely what the free market offers. And they are precisely why we need the blockchain.

APPENDIX: FAQ

Frequently asked questions.

GENERAL QUESTIONS

Are digital currencies money?

Yes. They are as valid as any other currency in circulation.

What gives digital cash its value?

The value of the digital cash comes from the same properties that give all money its value. It is fungible, recognizable, limited, private, liquid, costly to obtain, valued, useful for trade, useful as a store of value, and possession is ownership. If a digital currency is trusted, has good features, and holds its value, people will choose to use it.

Is digital cash as safe as my credit cards?

No. Digital cash is as safe as physical cash. Actually, it is somewhat safer because it can be backed up. Carrying and spending digital cash is somewhat safer than carrying and spending cash. But it is not as safe as credit cards.

It is possible for a bank to create a credit card for any digital currency. In that case, the answer is yes.

Is digital cash as safe as my debit cards?

It really depends on how you define safe. If you lose your debit card and your PIN is written on the back of it, your money isn't very safe.

Carrying and spending digital cash is somewhat safer than carrying and spending cash. But it is not as safe as monetary instruments backed by banks, such as debit cards. However, it is possible for banks to create debit cards for any digital currency. In that case, the answer is yes.

What happens if I lose my phone containing my digital cash?

Wallet software generally requires a PIN to access it. If your PIN is easy to figure out, you're in trouble when you lose your mobile device.

If you have backed up your digital wallet to your digital safe, you must quickly access your safe. You use your digital safe to invalidate the digital cash on your mobile device so that you can use the backup instead.

You must act fast to preserve your lost coins. If someone finds your phone and figures out your PIN before you can access the backup in your digital safe, it is possible for them to move your coins to their account. The blockchain will see the coins as no longer belonging to the account in your wallet. They will have successfully stolen your money.

If you have not backed up your wallet, then it is just like losing a real wallet with real cash. It's probably gone.

What happens if my phone dies? Do I lose my digital cash?

If you have backed up your wallet to your digital safe, you have not lost anything. You just transfer your coins in the backup to another account in your safe. The blockchain will see the coins as no longer belonging to the account in your wallet. When you get a new phone with wallet software, you can transfer the coins from your new account in your safe to your new wallet.

If you have not backed up your wallet, then it is just like losing a real wallet with real cash. It's gone.

What happens if many currencies in the world lose their value at the same time? Won't that affect digital currencies as well?

No. The value of other currencies does not necessarily affect the buying power of blockchain-based currencies.

If, for instance, the value of the US dollar collapsed, it would cause a domino effect that would destroy the value of most national currencies around the world. That is because the US dollar is used to back some 50-60 other national currencies. The current world economic system can't survive such an event.

However, it's very reasonable to expect that digital currencies could continue because they have no connection to the politics or finances of any particular nation. In a worldwide economic collapse, it is likely that digital cash would become *more* valuable, not less.

Isn't this a pretty subversive idea?

Bwahahahahahahahaaaaaaa!

Ok, the real answer is that it depends on your point of view. Blockchains are definitely a disruptive technology. History is filled with disruptive technologies (examples: mass produced cars, hydraulic backhoes, angioplasty, mainframe computers, personal computers, smartphones). It is normal that disruptive technologies should occur in a free market. Disruptive technologies always produce something better.

A free currency market vastly democratizes money and the economy. Everyone can be a player. Anyone can issue a currency and have it succeed if they can gain the trust of a large enough group of people and if their currency has features that people want.

Blockchains enable us to apply free market principles to money itself. This produces a more stable economy that is designed to maximize the benefit to all of its participants, not just a select few.

ECONOMICS

Why do you hate central banks so much?

I don't really hate them so much as see them as obsolete.

Central banks issue money by using debt. That is, all money gets into the economy through lending. Debt-based money requires the payment of interest, which concentrates wealth and power from the many to the few.

Central banks use central planning to control the economy. A few unelected people decide what interest rates will be, what unemployment levels will be, how easy or hard it is to get a loan, and so on. They control virtually all aspects of a nation's economy.

Blockchain-based currencies do not create or distribute money through debt. There is nothing in the system that concentrates money or power to any central authority.

Because blockchains are not run by the government, they are not subject to political expediency. No one can make blockchain-based currencies into a centrally planned economy.

THE BLOCKCHAIN AND GOVERNMENT

Digital currencies are seen in some countries as a foreign currency. Don't legal tender laws make it illegal to pay in foreign currencies?

No and yes. The answer varies by country. For example, legal tender laws in the US state that you must pay debts with US dollars *if* you incur the debt first. That is, if you eat a meal at a restaurant in the US, you are in debt to the restaurant and required to pay in US dollars.

On the other hand, if you go to a store to buy something and you and the store agree in advance to pay in something other than US dollars, you can.

Note that it is also possible to pay the restaurant owner in another currency *if* and *only if* both you and the owner agree on that in advance. So if the restaurant has a sign that says, "We Accept Bitcoins" you are free to pay in bitcoins and the restaurant owner is free to accept them.

It is also legal in the US to incur debts, take out loans, etc. in foreign currencies if you and the other party agree to do so in advance. As long as everyone involved knows what they're getting and agrees to it in advance, it's ok.

In some countries, it is illegal to buy, sell, or fulfill contracts with anything other than the national currency. However, the number of countries that have such laws is around half a dozen.

Some countries may see blockchain-based digital cash as a commodity (like gold) rather than a currency. These countries do not make it illegal to use commodities to buy and sell. But you may have to pay capital gains tax. Your wallet software generally is designed to handle this for you.

Won't the government see this as bad? Isn't it illegal to mint your own currency?

Some governments will definitely see blockchain-based digital currencies as bad, especially repressive regimes.

In most countries, however, they are perfectly legal because they are a foreign currency or just a commodity like gold or silver. Most countries do not prevent you from using them if both the buyer and the seller agree to do so in advance.

We already use many forms of private currency and currency-like products. Here's a partial list:

- Checks

- IOUs

- Discount coupon books sold for fundraising activities

- Discount coupons in newspaper ads

- Rebates on preferred customer cards

- Airline miles used to gain discounts on flights

- And many, many more.

Outlawing blockchain-based digital currencies means outlawing all of these as well. Virtual cash is deeply engrained in our economic system. Digital cash is really just a generalization of existing practices.

What if the government tries to shut down blockchains and digital currencies?

Nothing, that's what. As far as we know, digital cash stays well within the laws of all democratic governments. Free societies will be the place where the blockchain thrives. Digital currencies are by the people and for the people. They strengthen individual rights and empower everyone in the system.

Suppose one repressive government tries to outlaw digital currencies such as bitcoin. It won't make any difference at all. Blockchain-based systems are distributed worldwide. It is impossible for anyone to completely shut them down. Some countries, such as Russia, have outlawed this technology. It hasn't made a bit of difference, except to drive the technology into the hands of the Russian mafia. And they are very happy to ignore all government edicts and use blockchains anyway.

Even if a repressive government or a group of repressive governments does get an entire blockchain system shut down, it still won't matter. Almost all of the software is built on free and open source libraries. Anyone can use them to create their own digital currency system anywhere in the world. It really only takes one dedicated person to reconstitute a blockchain-based currency system. Going after any one person or group won't make any difference at all. You can even shut down the entire Internet and the system can survive using the telephone network, CB radios, or HAM radios.

THE BLOCKCHAIN AND BAD GUYS

Can't bad guys use bitcoins and other blockchain-based currencies to evade taxes, launder money, and stuff like that?

Yes. But they already do that with cash and by many other means. Anything of value can be used to accomplish those purposes. Bitcoins and similar currencies are no different in that respect. Therefore, they are no worse than what we already have. And in many ways, they are better.

Talking about the bad guys using digital cash is, in fact, deception by misdirection. The preferred currency of dictators, drug cartels, warlords, and bad guys around the world is the US dollar. These days, bad guys in certain places in South America, Asia, and Africa actually buy banks so that they can launder their ill-gotten gains[158]. They move their money across national borders disguised as legitimate bank transactions. If we say that the bitcoin or any other form of digital cash should be eradicated because bad guys use it then we must say the same of the US dollar.

Bitcoins in particular are not even a consideration to most criminals. They attract too much attention because government regulators are already looking at them. Instead, criminal organizations are already issuing their own digital currencies based on bitcoin technology. Western law enforcement reports that at least three such currencies are already in circulation in Russia and central Asia, and their acceptance is quietly growing around the world[159]. If

---

[158] For more information on the activities of organized crime using legitimate banks to launder money, please see How the Russian mafia use the banking system (especially Britain's) (http://thefinanser.co.uk/fsclub/2012/11/how-the-russian-mafia-use-the-banking-system-including-britains.html). For information on known members of organized crime that are owners or officers of mainstream banks, see http://rumafia.com.

[159] For information on digital currencies minted by organized crime, see Bye, Bitcoin: Criminals Seek Other Crypto Currency (http://www.darkreading.com/vulnerabilities-and-threats/bye-bitcoin-criminals-seek-other-

legitimate organizations don't get into this, the bad guys will. They won't care if countries outlaw it. They'll do it anyway.

And with the problems most national currencies are experiencing, people around the world will have profound motivations to accept and use digital currencies sponsored by criminal organizations as long as they're stable. It's actually possible that if we don't create a free market of competing currencies and our national currencies collapse, we could unintentionally hand over control of our entire worldwide economy to criminals. And no one could buy or sell unless they participated in the criminal economy. Hopefully we never see the conditions that could lead to such an extreme scenario. But applying free market principles to money itself ensures that this *can't* happen.

crypto-currency/d/d-id/1113864), and Mo Money Mo Problems (https://blogs.rsa.com/mo-money-mo-problems/).

ABOUT THE AUTHOR

David Conger is a computer scientist, author, former college professor, and self-taught economist. When he first encountered bitcoin and decided that it could never be more than an experiment, he made the mistake of wondering, "What would a *real* digital currency look like?" From that simple question came his six-year quest to design the Qbit.

David works at building the Qbit system and establishing the Qbit Federation. He seeks likeminded people to help him along the way and to contribute their innovative ideas to the Qbit system. He can be contacted through his company, Cognisaya, LLC (cognisaya.com), or through the Qbit Federation (qbitfederation.org).

www.ingramcontent.com/pod-product-compliance
Lightning Source LLC
Chambersburg PA
CBHW030841210326
41521CB00025B/448